1 MONTH OF
FREE
READING

at

www.ForgottenBooks.com

By purchasing this book you are eligible for one month membership to ForgottenBooks.com, giving you unlimited access to our entire collection of over 1,000,000 titles via our web site and mobile apps.

To claim your free month visit:
www.forgottenbooks.com/free1079017

ISBN 978-0-331-53135-0
PIBN 11079017

Sottom

||||| · ||||| · ||||| · ||||| · |||||

$$\frac{15}{45}\ \frac{2.5}{5}\ \frac{9}$$

8 3
3 0
—————
5 · 5 0
3 · 8 0
—————
1 0

5 · 3 0
3 · 8 0
—————
6 · 1 0

A

COMPLETE KEY

TO

SMILEY'S

NEW FEDERAL CALCULATOR,

OR

Scholar's Assistant;

IN WHICH THE

METHOD OF SOLVING ALL THE QUESTIONS CONTAINED
IN THAT WORK IS EXHIBITED AT LARGE.

DESIGNED

TO FACILITATE THE LABOUR OF TEACHERS, AND ASSIST
THOSE WHO HAVE NOT THE ADVANTAGE OF A
TUTOR'S AID.

BY THOMAS T. SMILEY,
TEACHER.

Author of an Easy Introduction to the Study of Geography. Also,
of Sacred Geography, for the use of Schools.

Philadelphia:

PUBLISHED AND FOR SALE BY GRIGG & ELLIOTT, No. 9,
NORTH 4th St. AND FOR SALE BY BOOKSELLERS AND
COUNTRY MERCHANTS GENERALLY IN THE
SOUTHERN AND WESTERN STATES.

1835.

CONTENTS.

———

EXPLANATION OF CHARACTERS.

Signs.	Significations.
$=$	Equal; as $20s. = £1.$
$+$	Addition, (or more) as $6+2=8.$
$-$	Subtraction, (or less) as $8-2=6.$
$\times$	Multiplication, (or multiplied by) as $6\times2=12.$
$\div$	Division, (or divided by) as $6\div2=3.$
$: :: :$	Proportionally; as $2:4::6:12.$
$\sqrt{}$ or $\sqrt[2]{}$	Square Root: as $\sqrt[2]{64}=8.$
$\sqrt[3]{}$	Cube Root; as $\sqrt[3]{64}=4.$
————	A vinculum; denoting the several quantities over which it is drawn, to be considered jointly as a simple quantity.

A KEY

The New Federal Calculator.

—◦●◦—

SIMPLE ADDITION.

EXAMPLES.

(8)	4829	(9)	91769	(10)	876994
	1234		14678		213678
	6101		80032		482906
	3014		71897		809769
	5618		76989		376980
	20796		335365		2760327

(11)	389261	(12)	2136784	(13)	3769694
	789794		8297698		4976082
	849798		8297694		4569761
	487697		4897695		8213243
	999996		1234697		4876962
	948219		7092032		4876920
	4464765		31956600		31282662

(14)	37856	(15)	378269	(16)	141
	975		402607		5672
	1234		702		82971
	14		1246		34676
	5612		2132		1450
	2075		45178		427
	16287		10276		12
	64053		840410		125358

SIMPLE ADDITION.

(17)	(18)	(19)	(20)
14	36	3797	205
16	97	95	20
23	125	2	840
29	384	75	970
80	1176	876	367
31	———	9750	1001
100	1818	———	———
———		14595	3403
293			

(21)	(22)	(23)
365	300	75960800
807	75	225000
560	2	140
25	47	———
37	33	76185940
101	9784	
———	20150	
1895	765091	
	1075047	
	———	
	1870529	

PRACTICAL EXERCISES.

(24)	(25)	(26) $	(27) $	(28) Miles.
35	275	30	50	37
21	196	12	25	33
——	——	5	125	40
56	471	——	216	35
		$47	——	——
			416	145

(29) Sheep.	(30)	(31)	(32) bar.	$
A's 34	25	8	400 for	2000
B's 47	15	15	550	2750
C's 54	40	19	——	——
——	9	12	950	$4750
135	——	——		
	89	54		

MULTIPLICATION.

CASE I.

(8) 3948769768
    3
    _____
    11846309304

(9) 87051298
    4
    _____
    348205192

(10) 976201696769
     5
     _____
     4881008493845

(11) 456978426976
     6
     _____
     2741870561856

(12) 8079696769
     7
     _____
     56560891383

(13) 97696429769
     8
     _____
     781587438152

(14) 28769842369
     9
     _____
     258928581321

(15) 7696297694780
     10
     _____
     76962976947800

(16) 56976969976845
     11
     _____
     626746887452955

(17) 7029676956
     12
     _____
     84356523472

(18) 84976876969
     12
     _____
     1019722523628

(19) 90216814096771
     12
     _____
     1082601769160052

(20) 4218
     2
     ____
     8436

(21) 7321
     3
     ____
     21963

(22) 87692
     4
     ____
     350768

(23) 95698
     5
     ____
     478490

(24) 10691
     6
     ____
     64146

(25) 31078
     7
     ____
     217546

(26) 109019
     8
     ____
     872152

(27) 900078
     9
     ____
     8100702

(28)   826870
             10
      8268700

(29)   278976
            11
      3068736

(30)   12569769
            12
   150837228

CASE 2.

EXAMPLES.

(34) 39786948
          197

  278508636
  358082532
   39786948

  7838028756

(35) 4978829
         408

  39830632
  19915316

 2031362232

(36) 8735698
        5706

  52414188
  61149886
  43678490

 49845892788

(37) 84016978
         3761

   84016978
  504101868
  588118846
 252050934

 315987854258

(38) 49569876
         4817

  346989132
   49569876
  396559008
  198279504

 238778092692

(39) 9637842
       9078

  77102736
  67464894
  86740578

 87492329676

(40) 9786
      13

  29358
   9786

 127218

(41) 8475
     29

  76275
  16950

 245775

(42) 11271
      35

  56355
  33813

 394485

(43) 19004
          305
      _____
       95020
      57012
      _____
      5796220
      _____

(44) 76976
        489
      _____
       692784
       615808
       307904
      _____
       37641264
      _____

(45) 84769
        976
      _____
       508614
       593383
       762921
      _____
       82734544
      _____

(46)   1978987
          4809
      _____
       17810883
       15831896
        7915948
      _____
       9516948483
      _____

(47)   9807094
          5047
      _____
        68649658
        39228376
        49035470
      _____
        49496403418
      _____

CASE 3.

EXAMPLES.

(48) 37|00
       2|00
      _____
      740000
      _____

(49) 4870
        25|00
      _____
       24350
        9740
      _____
       12175000
      _____

(50) 4087|00
       906|000
      _____
        24522
        36783
      _____
       370282200000
      _____

(51)   876956
          99|0000
      _____
       7892604
       7892604
      _____
       868186440000
      _____

CASE 4.

EXAMPLES.

(53) 8976	(54) 7696	(55) 87698	(56) 20784
6	9	9	12
53856	69264	789282	249408
8	9	8	9
430848	623376	6314256	2244672

(57) 81207	(58) 47696	(59) 75687	(60) 34075
11	12	7	6
893277	572352	529809	204450
12	12	8	6
10719324	6868224	4238472	1226700

PRACTICAL EXERCISES.

(61) $25	(62) 15	(63) $250	(64) $150
5	4	7	4
$125	60	$1750	$600

(65) $100	Or thus, 100		(66) 18175
25	5		14
500	500		72700
200	5		18175
$2500	$2500		254450

SUBTRACTION.

EXAMPLES.

(4) 859768	(5) 9076048	(6) 532147878
124978	7940689	139876956
734790	1135359	392270922

(7) 100000
84321

15679

(8) 75381478
39040217

36341261

(9) 102070845
19768799

82302046

(10) 196
37

159

(11) 487
96

391

(12) 875
302

573

(13) 967
351

616

(14) 1001
487

514

(15) 9765
1307

8458

(16) 87696
10091

77605

(17) 455692
300120

155572

(18) 1000000
1

999999

PRACTICAL EXERCISES.

(19) 25
8
--
17
--

(20) 75
42
--
33
--

(21) 7896
4389

3507
----

(22) 4875
2976

1899
----

(23) 1240 375
1082 567
---- 140
$158 ----
Sum 1082

(24) 5487
2075

3412

325
750
1000

2075 Sum.

(25) 25 containing 250
9   75
--   ----
16   175
--   ----

DIVISION.

EXAMPLES OF SHORT DIVISION.

(7) 2)56789768

    28394884

(8) 3)37297687769

    1243256256+1

(9) 4)469769876

    117442469

DIVISION.

(10) 5)849768769
      169953753+4

(11) 6)756976874
      126162812+2

(12) 7)87694213628
      12527744804

(13) 8)80269687
      10033710+7

(14) 9)376948769
      41883196+5

(15) 11)876956788
      79723344+4

(16) 12)4976876946782
      414739745565+2

(17) 12)89769762048769
      7480813504064+1

(18) 2)3976
      1988

(19) 3)8769
      2923

(20) 4)47876
      11969

(21) 5)8767
      1753+2

(22) 6)9698
      1616+2

(23) 7)97899
      13985+4

(24) 8)80409
      10051+1

(25) 9)981021
      109002+3

(26) 10)897697
      89769+7

(27) 11)9876978
      897907+1

28) 12)4967844
      413987

PRACTICAL EXERCISES.

(29) 2)12
    6

(30) 7)350
    50

(31) 8)8736
    4)1092
    273

(32) 3)3966
    1322

LONG DIVISION.

EXAMPLES.

(35) 13)875(67
         78
      ——
         95
         91
      ——
          4
      ——

(36) 15)476(31
         45
      ——
         26
         15
      ——
         11
      ——

(37) 18)958(53
         90
      ——
         58
         54
      ——
          4
      ——

(38) 28)1475(52
         140
      ——
          75
          56
      ——
          19
      ——

(39) 31)4277(137
         31
      ——
         117
          93
      ——
         247
         217
      ——
          30
      ——

(40) 37)25757(696
         222
      ——
         355
         333
      ——
         227
         222
      ——
           5
      ——

(41) 41)256976(6267
         246
      ——
         109
          82
      ——
         277
         246
      ——
         316
         287
      ——
          29
      ——

(42) 48)337979(7041
         336
      ——
         197
         192
      ——
          59
          48
      ——
          11
          —

(43) 59)997816(16912·
      59
      ___
      407
      354
      ___
      538
      531
      ___
       71
       59
      ___
      126
     ·118
      ___
        8
        —

(44) 98)999987695(10203956
      98
      ___
      199
      196 -
      ___
      387
      294
      ___
      936
      882
      ___
      549
      490
      ___
      595
      588
      ___
        7

(45) 125)4697680424(37581443
      375
      ___
      947
      875
      ___
      726
      625
      ___
      1018
      1000
      ___
      180
      125
      ___
      554
      500
      ___
      542
      500
      ___
      424
      375
      ___
       49

(46) 396)387690204886(979015668
      3564
      ____
      3129
      2772
      ____
      3570
      3564
      ____
      620
      396
      ____
      2244
      1980
      ____
      2648
      2376
      ____
      2723
      2376
      ____
      3526
      3168
      ____
       358

(7)　876)4876620048769(5566232932
4380

4960
4380

5802
5256

5460
5256

2040
1752

2884
2628

2568
1752

8167
7884

2836
2628

2089
1752

337

(8)　1478)8769626000402(5933576454
7390

13798
13302

4962
4434

5286
4434

8520
7390

11300
10346

9540
8868

6724
5912

8120
7390

7302
5912

1390

(49) 87696)9876976876872O497(1126274501　　(50) 97680|0000)8976478976|0000(91896

```
        87696                                87912

        110737                               18527
        87696                                9768

        230416                               87598
        175392                               78144

         550248                              94549
         526176                              87912

          240727                             66377
          175392                             58608

           653352                            77696
           613872

            394800
            350784

             440164
             438480

              168497
              87696

               80801
```

(51)　　　14769Q0|00000)4789768214|00000(3242 Ans.
```
                4430940

                 3588282
                 2953960

                  6343221
                  5907920

                   4353014
                   2953960

             Rem. 1399054
```

(52)   45)9847(218
        90

        84
        45

        397
        360

        Rem. 37

(53)   391)1259678(3221
        1173

        866
        782

        847
        782

        658
        391

        Rem. 267

(54)   148)225476(1523
        148

        774
        740

        347
        296

        516
        444

        Rem. 72

(55)   25)375(15 bushels.
        25

        125
        125

```
(56)  75)87735840(1169811          (57)   49850)99700(2
      75                                         99700
      ───
      127
       75
      ───
      523
      450
      ───
      735
      675
      ───
      608
      600
      ───
       84
       75
      ───
       90
       75
      ───
       15 Rem.
```

When the divisor is the exact product of any two figures multiplied together.

EXAMPLES.

```
(61)  5)9756                  (62)  9)8491
      ─────────                     ─────────
      7)1951+1 1st Rem.             9)943+4
      ───────────                   ─────────────────
      278+5 2d Rem.                 104+7×9+4=67
        ×5                          ─────────────────
      ─────────
      25+1=26
      ─────────
```

```
(63)  9)44767                 (64)  7)92017
      ─────────                     ─────────
      2)4974+1 Rem.                 8)13145+2
      ─────────                     ─────────────  Rem.
      2487                          1643+1×7+2=9
      ─────────
```

(65) 11)55210

9)5019+1

Rem.
557+6×11+1=67

(66) 6)38751

8)6458+3

Rem.
807+2×6+3=15

(67) 12)99876

9)8323

Rem.
924+7×12=84

(68) 12)37967

12)3163+11

Rem.
263+7×12+11=95

PRACTICAL EXERCISES.

(69) 5)3775

5)755

Ans. 151

(70) 12)480

8)40

Ans. 5 lb

(71) 12)14400

12)1200

Ans. 100

(72) 12)1800

6)150

Ans. 25

(73) 12)396

11)33

Ans. $3

EXAMPLES IN ADDITION, MULTIPLICATION, SUBTRACTION
AND DIVISION.

(1) 50
     50

     100
     —25
     ---
     75 Ans.

(2) 40 10
     20 10
     2)20 20

     Ans. 10

(3) 25000
     13000
     2)12000

     $6000

(4) Bought 8200   Sold 3756   (5) 50)2450(49 miles. Ans.
         5060   4879   200
        ─────   ─────   ────
        13200   8635   450
         8635   450
        ─────   ─────   ────
    Ans. 4565

(6)   Bought 24 bags, containing 3000 ℔
           Sold   15   1736
                    ──   ────
           Remains 9 bags, containing 1264 ℔

(7) Days 365)2920(8 dols. per day.   Yearly income 2920
               2920   Spends yearly 1769
              ─────   ────
                                     Saves per year $1151

COMPOUND ADDITION.

FEDERAL MONEY.

EXAMPLES.

	$	cts.	m.			$	cts.			$	cts.
(2)	46	75	5		(3)	37	$68\frac{3}{4}$		(4)	72	$62\frac{1}{4}$
	79	37	8			95	$37\frac{1}{4}$			85	$87\frac{1}{2}$
	43	50	0			43	25			20	$12\frac{1}{4}$
	97	37	5			79	$56\frac{1}{4}$			45	$18\frac{3}{4}$
										94	$37\frac{1}{2}$
	$267	00	8			$255	$87\frac{1}{2}$			42	$68\frac{3}{4}$
										79	$18\frac{3}{4}$
										$440	$06\frac{1}{4}$

(5)

$	cts.
54	75
37	$37\frac{1}{2}$
93	$18\frac{3}{4}$
149	$87\frac{1}{2}$
503	$68\frac{3}{4}$
979	$12\frac{1}{2}$
2194	$18\frac{3}{4}$
$4012	$18\frac{3}{4}$

(6)

$	cts.
29	25
34	$37\frac{1}{2}$
188	$68\frac{3}{4}$
265	$12\frac{1}{2}$
1783	$18\frac{3}{4}$
8579	$56\frac{1}{4}$
6	$87\frac{1}{2}$
$10887	$06\frac{1}{4}$

(7)

$	cts
1	$18\frac{3}{4}$
2	50
	$87\frac{1}{2}$
	$93\frac{3}{4}$
1	$87\frac{1}{2}$
2	$68\frac{3}{4}$
	$37\frac{1}{2}$
	$87\frac{1}{2}$
1	$93\frac{3}{4}$
$13	25

(8)

$	cts.
5	00
18	50
8	$87\frac{1}{2}$
1	$18\frac{3}{4}$
14	50
0	$87\frac{1}{2}$
5	$37\frac{1}{2}$
7	$87\frac{1}{2}$
20	00
$82	$18\frac{3}{4}$

(9)

$	cts.
1	$87\frac{1}{2}$
1	$68\frac{3}{4}$
0	$43\frac{3}{4}$
1	$37\frac{1}{2}$
0	$93\frac{3}{4}$
0	$56\frac{1}{4}$
0	$37\frac{1}{2}$
0	$31\frac{1}{4}$
0	$12\frac{1}{2}$
$7	$68\frac{3}{4}$

STERLING MONEY.

EXAMPLES.

(2)

	£	s.	d.
	7	9	$4\frac{1}{2}$
	13	7	$6\frac{3}{4}$
	4	5	2
	10	18	$10\frac{3}{4}$
Ans.	36	1	0

(3)

	£	s.	d.
	4	6	4
	47	19	7
	159	5	3
	78	6	$11\frac{3}{4}$
Ans.	289	18	$1\frac{3}{4}$

(4)

	£	s.	d.
	565	3	7
	382	13	5
	592	9	2
	856	17	3
	259	9	8
Ans.	2656	13	1

	£	s.	d.
(5)	142	16	7
	489	3	4
	726	15	9
	573	4	8
	628	12	6
Ans.	2560	12	10

	£	s.	d.
(6)	763	7	4
	39	4	9
	162	17	2
	459	15	0
	473	12	8
Ans.	1898	16	11

	£	s.	d.
(7)	69	18	7
	175	2	6
	1582	19	4
	175	13	9
	143	13	8
	212	0	7
Ans.	2359	8	5

	£	s.	d.
(8)	1776	12	8
	412	16	5
	369	7	2
	469	15	10
	573	19	2
	1987	14	8
	4823	15	11
Ans,	10414	1	10

	£	s.	d.
(9)	985	4	9
	186	13	4
	1569	18	4
	183	0	8
	0	17	4
	0	0	7
Ans.	2925	15	0

AVOIRDUPOIS WEIGHT.

	T.	cwt.	qr.	lb.	oz.	dr.
(2)	7	11	2	16	4	13
	15	7	3	8	16	7
	138	19	1	12	8	13
	42	8	3	19	12	4
	357	6	2	8	3	3
Ans.	561	14	1	7	13	8

	T.	cwt.	qr.	lbs.	oz.	dr.
(3)	12	16	1	19	15	0
	114	10	2	12	4	13
	72	4	2	24	14	3
	176	15	3	4	15	11
Ans.	376	7	2	6	1	11

	T.	cwt.	qr.	lb.	oz.	dr.
(4)	139	19	3	18	13	10
	1754	10	2	11	2	14
	27	3	0	14	11	0
	0	13	0	0	13	0
Ans.	1922	6	2	17	8	8

TROY WEIGHT.

lbs.	oz.	dwts.	gr.		lbs.	oz.	dwts.	gr.		lbs.	oz.	dwts.	gr.
(2) 185	2	19	20	(3)	16	4	18	6	(4)	172	11	19	22
56	9	15	6		7	9	11	22		12	4	13	12
1472	11	2	17		163	7	12	18		18	5	11	20
385	0	8	5		17	0	13	0		119	11	18	18
10	8	7	12							0	0	2	13
				Ans.	204	10	15	22		0	10	0	20
Ans. 2110	8	13	12										
									Ans.	324	8	2	9

APOTHECARIES' WEIGHT.

℔	℥	ʒ	Ɵ	gr.		℔	℥	ʒ	Ɵ	gr.		℔	℥	ʒ	Ɵ	gr.
(2) 84	7	6	0	12	(3)	18	0	1	0	12	(4)	182	3	1	0	0
132	5	0	2	0		175	10	5	0	10		12	1	0	2	17
16	2	2	2	8		472	3	1	2	3		17	2	4	2	15
1427	6	7	0	19		0	11	7	2	0		0	10	2	1	19
14	0	6	1	9												
					Ans. 667	1	7	2	5	An. 212	5	1	1	11
A. 1674	10	7	1	8												

LONG MEASURE.

	yd.	ft.	in.		L.	m.	f	p	yd.	ft.	in.		L.	m.	f.	p	yd.	ft.	in.
(2)	3	2	11	(3)	172	2	3	19	2	2	4	(4)	462	1	7	29	1	1	10
	1	1	9		0	0	0	14	1	0	3		0	0	0	11	0	1	10
	2	0	8		0	1	2	29	0	0	10		4	1	2	28	1	2	9
	3	1	10		0	0	4	0	0	0	0		0	0	0	13	0	0	0
	2	0	4		0	0	2	0	0	0	10								
	6	2	7		0	0	0	0	3	2	3	Ans.	467	0	3	1	4	0	5
Ans. 20	1	1		Ans.	173	1	4	23	2½	0	6								

CLOTH MEASURE.

	E. E.	qr.	n.		E. F.	qr.	n.
(2)	72	3	2	(3)	19	2	3
	536	2	1		728	1	2
	847	1	3		142	0	1
	1453	0	2		816	0	0
	41	2	0		32	1	2
Ans.	2951	0	0	Ans.	1739	0	0

	yd.	qr.	na.			E. Fr.	qr.	na.
(4)	19	2	3		(5)	143	0	3
	14	2	0			17	2	2
	32	0	2			172	1	1
	0	3	1			182	1	3
	142	3	2			132	8	2
						72	1	1
Ans.	210	0	0					
					Ans.	720	1	0

LAND MEASURE.

	A.	R.	P.			A.	R.	P.			A.	R.	P.
(2)	487	2	17		(3)	22	2	0		(4)	132	3	25
	25	3	28			700	3	27			654	0	17
	67	0	32			47	0	5			462	3	25
	45	1	16			39	0	0			16	0	4
	26	0	29			47	2	39			1665	3	38
						0	3	28					
Ans.	652	1	2							Ans.	2931	3	29
					Ans.	858	0	19					

LIQUID MEASURE.

	hhd	gal.	qt.	pt.		T.	h.	gal.	qt.	pt.		T.	h.	gal.	qt.	pt.
(2)	385	42	3	1		(3)	19	2	19	0	0		(4) 862	1	0	1	0
	27	36	2	0			45	0	0	1	1		0	0	32	0	1
	132	17	0	0			0	3	17	2	0		0	0	37	2	0
	729	25	0	0			0	0	21	0	1		0	0	32	1	0
	163	47	2	1									0	2	0	0	1
						Ans.	65	1	58	0	0						
Ans.	1438	43	0	0								Ans.	863	0	39	1	0

DRY MEASURE.

	B.	p.	qt.	pt.		B.	p.	qt.	pt.		B.	p.	qt.	pt.
(2)	47	2	4	1		(3)	754	2	5	0		(4)	144	3	2	1
	635	0	3	0			469	0	2	0			0	1	2	0
	247	3	0	1			385	2	7	1			0	0	3	1
	285	0	2	0			375	0	0	1			462	3	0	1
	734	2	5	0			0	3	2	0			72	0	5	1
Ans.	1950	0	7	0		Ans.	1985	1	1	0		Ans.	680	0	6	0

TIME.

	Y.	m.	w.	d.	h.	m.	sec.
(3)	172	0	1	0	4	0	52
	0	0	0	0	0	34	18
	15	4	0	5	3	27	0
	0	0	1	3	21	35	18
Ans.	187	4	3	2	5	37	28

	Y.	m.	w.	d.	h.	m.	sec.
(4)	462	4	0	0	5	37	24
	62	0	0	0	11	0	24
	0	0	1	5	0	13	0
	0	6	1	4	13	12	37
Ans.	524	10	3	3	6	3	25

MOTION, OR CIRCLE MEASURE.

	sig.	°	'	"
(2)	2	7	32	16
	0	5	27	24
	1	6	17	13
	0	7	38	24
	4	5	42	19
Ans.	8	2	37	36

	sig.	°	'	"
(3)	5	10	46	38
	0	11	37	18
	1	0	47	12
	0	0	0	18
	2	0	0	52
	1	15	12	23
	0	11	57	29
Ans.	10	20	22	10

	sig.	°	'	"
(4)	0	0	45	0
	1	9	0	18
	0	14	21	34
	2	8	13	54
	4	7	12	19
	0	0	47	32
Ans.	8	10	20	37

APPLICATION.

	$	cts.
(1)	375	45
	142	$37\frac{1}{2}$
	1375	$56\frac{1}{4}$
Ans.	1893	$38\frac{3}{4}$

	Y.	qr.	na.
(2)	57	2	0
	29	3	2
	45	1	0
	32	3	1
	38	2	0
	38	2	0
Ans.	242	1	3

	B.	p.	qt.
(3)	2	2	0
	3	3	5
	3	1	1
	2	0	4
Ans.	11	3	2

	A.	R.	P.
(4)	142	2	0
	32	3	12
	108	3	18
Ans.	284	0	30

	Y.	qr.	na.
(5)	15	3	0
	18	1	2
	25	3	2
Ans.	60	0	0

	M. fur. p.		
(6)	43	3	0
	29	0	34
	57	2	32
	12	3	18
Ans.	142	2	4

	B.	p.	qt.
(7)	756	2	0
	756	2	0
	756	2	0
	854	0	5
	854	0	5
Ans.	3977	3	2

COMPOUND MULTIPLICATION.

EXAMPLES.

FEDERAL MONEY.

	$	cts.
(4)	26	$18\frac{3}{4}$
		6
Ans.	157	$12\frac{1}{2}$

	$	cts.	m.
(5)	100	40	4
			10
Ans.	1004	04	0

	$	cts.
(6)	56	$18\frac{3}{4}$
		9
Ans.	505	$68\frac{3}{4}$

	$	cts.	m.
(7)	25	37	5
			8
Ans.	203	00	0

	$	cts.
(8)	565	$62\frac{1}{2}$
		12
Ans.	6787	50

ENGLISH MONEY.

	£	s.	d.
(2)	14	6	$0\frac{1}{4}$
			9
Ans.	128	14	$2\frac{1}{4}$

	£	s.	d.
(3)	111	11	$\frac{1}{2}$
			10
Ans.	1115	18	9

	£	s.	d.
(4)	37	6	9½
			5
Ans.	186	13	11½

	£	s.	d.
(5)	56	8	7¾
			9
Ans.	507	17	9¾

AVOIRDUPOIS WEIGHT.

	T.cwt.	qr.	lb.	oz.	dr.
(2)	6 14	2	7	5	2
					4
Ans.	26 18	1	1	4	8

	qr.	lb.	oz.	dr.
(3)	3	16	7	8
				10
Ans.	35	24	11	0

	Cwt.	qr.	lb.
(4)	1	2	6
			10
Ans.	15	2	4

	Cwt.	qr.	lb.
(5)	4	3	17
			11
Ans.	53	3	19

TROY WEIGHT.

	lb.	oz.	dwt.	gr.
(2)	43	0	8	10
				4
Ans.	172	1	13	16

	lb.	oz.	dwt.	gr.
(3)	113	6	0	6
				6
Ans.	681	0	1	12

	lb.	oz.	dwt.
(4)	17	9	14
			10
Ans.	178	1	0

	lbs.	oz.	dwt.	gr.
(5)	41	6	18	2
				7
Ans.	291	0	6	14

	lbs.	oz.	dwt.	gr.
(6)	91	4	14	16
				8
Ans.	731	1	17	8

APOTHECARIES' WEIGHT.

	℔	℥	ʒ	Ɗ	gr.
(2)	53	10	0	2	12
					9
Ans.	484	6	7	2	8

	℔	℥	ʒ	Ɗ	gr.
(3)	17	5	6	1	4
					12
Ans.	209	9	4	2	8

	℔	℥	ʒ	℈
(4)	76	4	1	2
				9
Ans.	687	1	7	0 -

	℔	℥	ʒ	℈	gr.
(5)	95	1	2	1	11
					11
Ans.	1046	2	3	2	1

LONG MEASURE.

	L.	M.	fur.	p.
(2)	4	2	2	29
				7
Ans.	33	1	3	3

	M.	fur.	p.	yd.	ft.	in.
(3)	18	3	20	1	2	10
						5
Ans.	92	1	21	3½	2	2

	Deg.	m.	fur.
(4)	6	40	7
			10
Ans.	66	48	6 .

	M.	fur.	p.
(5)	44	6	20
			7
Ans.	313	5	20

CLOTH MEASURE.

	E.E.	qr.	na.
(2)	37	4	2
			8
Ans.	63	1	0

	E.Fl.	qr.	na.
(3)	18	0	3
			12
Ans.	217	4	0

	E.Fr.	qr.	na.
(4)	14	1	3
			9
Ans.	129	0	3

	Yds.	qr.	na.
(5)	19	1	2
			5
Ans.	96	3	2

	E. E.	qr.
(6)	56	3
		9
Ans.	509	2

LAND MEASURE.

	A.	R.	P.
(2)	19	3	20
			6
Ans.	119	1	00

	A.	R.	P.
(3)	10	0	33
			9
Ans.	91	3	17

	A.	R.	P.
(4)	1	3	11
			10
Ans.	18	0	30

	A.	R.	P.
(5)	63	3	18
			11
Ans.	702	1	38

LIQUID MEASURE.

(2)
```
    T. hhd. gal. qt. pt.
    1  2  16  3  1
                10
```
Ans. 15 2 42 3 0

(3)
```
    P. hhd. gal. qt. pt.
    4  1  19  3  1
                5
```
Ans. 23 0 36 1 1

(4)
```
    T. h. gal. qt.
    3  2  50  2
            8
```
Ans. 29 2 26 0

(5)
```
    H. gal. q. pt.
    4  41  0  1
            10
```
Ans. 46 33 1 0

DRY MEASURE.

(2)
```
    Bu. pe. qt. pt.
    1  3  3  2
          4
```
Ans. 7 2 0 0

(3).
```
    Bu. pe. qt. pt.
    110  3  0  2
            4
```
Ans. 443 0 4 0

(4)
```
    B. pe. qt. pt.
    44  0  0  1
          7
```
Ans. 308 0 3 1

(5)
```
    P. qt.
    3  1
       9
```
Ans. Bush. 7 0 1

TIME.

(2)
```
    Y.  m. w. d. h. min. sec.
    17  8  2  6  4  40  18
                        6
```
Ans. 106 0 1 2 4   1 48

(3)
```
    W. d.  h.
    3  5  22
          12
```
Ans. 46 1 0

(4)
```
    Y. m. w. d.
    7  0  4  4
             9
```
Ans. 63 10 1 1

(5)
```
    Y. m. w. d.
    15  2  0  6
             8
```
Ans. 121 4 2 6

RULE 2.

EXAMPLES.

(2) Multiply 37 10 6¾ by 48   (3). 66 37 5 by 36

£ s. d. $ cts. m.

6×8=48   6×6=36

225	3	4½
		8

Ans. 1801 7 0

398	25	0
		6

Ans. 2389 50 0

(4) $ cts. m.   (5) $ cts.

44 25 3 by 56   12 18¾ by 96

7×8=56   12×8=96

309	77	1
		8

Ans. 2478 16 8

146	25
	8

Ans. 1170 00

(6)   £ s. d.   (7) £ s. d.

45 6 9½ by 120   96 12 3¾ by 144

12×10=120   12×12=144

544	1	6
		10

Ans. 5440 15 0

1159	7	9
		12

Ans. 13912 13 0

(8)   A. R. P.   (9) M. f. p.

47 3 20 by 54   48 7 25 by 88

6×9=54   11×8=88

287	1	0
		9

Ans. 2585 1 0

538	3	35
		8

Ans. 4307 7 0

$$\begin{array}{ccc} lb. & oz. & dr. \end{array}$$
(10)   56   9   6 by 84

$$12 \times 7 = 84$$

681   9   0
          7

Ans. 4772   3   0

RULE 3.

(2) Multiply \$ 7   cts. $87\frac{1}{2}$

$$11 \times 4 + 3 = 47$$

86   $62\frac{1}{2}$
      4

346   50
23   $62\frac{1}{2}$

Ans. 370 $12\frac{1}{2}$

(3)   \$ 28   cts. $68\frac{3}{4}$

$$11 \times 6 + 2 = 68$$

315   $56\frac{1}{4}$
        6

1893 $37\frac{1}{2}$
57   $37\frac{1}{2}$

Ans. 1950   75

(4)   \$ 49   cts. 75×3
        12

597   00
        7

4179   00
149   25

Ans. 4328   25

(5)   \$ 94   cts. $18\frac{3}{4} \times 1$
        10

941   $87\frac{1}{2}$
        3

2825 $62\frac{1}{2}$
94   $18\frac{3}{4}$

Ans. 2919 $81\frac{1}{4}$

(6)

$ cts.

42 31¼ × 3
      11
_____

465 43¾
     5
_____

2327 18¾
126 93¾
_____

Ans. 2454 12½

(7)

£ s. d.

28 7 6½ × 1
       4
_____

113 10 2
      7
_____

794 11 2
 28 7 6½
_____

Ans. 822 18 8½

(8)

£ s. d.

34 8 4¾ × 1
     11
_____

378 12 4¼
     6
_____

2271 14 1½
  34 8 4¾
_____

Ans. 2306 2 6¼

(9)

Cwt. qr. lb.

7 3 22 × 1
     10
_____

79 1 24
    5
_____

397 1 8
  7 3 22
_____

Ans. 405 1 2

(10)

lbs. oz. dwts.

12 5 8 × 3
    12
_____

149 4 16
     3
_____

448 2 8
37 4 4
_____

Ans. 485 6 12

(11)

M. f. p.

4 6 21 × 3
    12
_____

57 6 12
     7
_____

404 4 4
 14 3 23
_____

Ans. 418 7 27

RULE 4.

	$	cts.
(2) Multiply	1	56½×6
		10

15	65×5
	10

156	50
	4

626	00
78	25
9	39

Ans. 713 64

	$	cts.
(3)	2	87½×6
		10

28	75×7
	10

287	50
	5

1437	50
201	25
17	25

Ans. 1656 00

	$	cts.
(4)	4	31¼×9
		10

43	12½×7
	10

431	25
	6

2587	50
301	86½
38	81¼

Ans. 2928 18¾

	$	cts.
(5)	18	93¾×7
		10

189	37½×5
	10

1893	75
	4

7575	00
946	87½
132	56¼

Ans. 8654 43¾

	$	cts.
(6)	25	$43\frac{3}{4} \times 9$
		10
	254	$37\frac{1}{2} \times 7$
		10
	2543	75
		8
	20350	00
	1780	$62\frac{1}{2}$
	228	$93\frac{3}{4}$
Ans.	22359	$56\frac{1}{4}$

	$	cts.
(7)	0	$1\frac{3}{4} \times 6$
		10
	0	$17\frac{1}{2} \times 6$
		10
	1	75×2
		10
	17	50
		2
	35	00
	3	50
	1	00
		$10\frac{1}{2}$
Ans.	39	$65\frac{1}{2}$

	$	cts.
(8)	10	$16\frac{1}{2} \times 9$
		10
	101	65×3
		10
	1016	50
		9
	9148	50
	304	95
	91	$48\frac{1}{2}$
Ans.	9544	$93\frac{1}{2}$

	£	s.	d.
(9)	37	18	$6\frac{1}{4} \times 5$
			10
	379	5	$2\frac{1}{2} \times 7$
			10
	3792	12	1
			3
	11377	16	3
	2654	16	$5\frac{1}{2}$
	189	12	$7\frac{1}{4}$
Ans.	14222	5	$3\frac{3}{4}$

	£.	s.	d.
(10)	48	14	$2\frac{1}{2}\times9$
			10
	487	2	1×8
			10
	4871	0	10
			4
	19484	3	4
	3896	16	8
	438	7	$10\frac{1}{2}$
Ans.	23819	7	$10\frac{1}{2}$

	£.	s.	d.
(11)	64	2	8×5
			10
	641	6	6×5
			10
	6413	6	6
			5
	32066	13	4
	3206	13	4
	320	13	4
Ans.	35594	0	0

	£.	s.	d.
(12)	58	9	$6\frac{3}{4}\times6$
			10
	584	15	$7\frac{1}{2}\times9$
			10
	5847	16	3
			3
	17543	8	9
	5263	0	$7\frac{1}{2}$
	350	17	$4\frac{1}{2}$
Ans.	23157	6	9

	M.	f.	p.
(13)	25	3	18×5
			10
	254	2	20×6
			10
	2543	1	0×2
			10
	25430	10	0
	5086	2	0
	1525	7	0
	127	1	10
Ans.	32170	4	10

	F. in.b.c.		Yd. qr.n.		Hhd.gal.qt.
(14)	48 4 2 × 7	(15)	22 2 1 × 4	(16)	4 37 2 by 4250
	10		10		10

483 10 2 × 8	225 2 2	45 60 0 × 5
10	10	10

4838 10 2 × 5	2256 1 0 × 2	459 33 0 × 2
10	10	10

48388 10 2	22562 2 0	4595 15 0
2	3	4

96777 9 1	67687 2 0	18380 60 0
24194 5 1	4512 2 0	919 3 0
3871 1 1	90 1 0	229 48 0
338 8 2		

Ans. 72290 1 0   Ans. 19529 48 0

Ans. 125182 0 2

APPLICATION.

(1) $12.50	(2) $1.07	(3) $5.62¼	(4) $1.12½
5	9	12	6
Ans. 62.50	Ans. 9.63	Ans. 67.47	6.75
			4
			Ans. 27.00

	£ s. d.		$ cts.
(5)	0 2 2 by 63	(6)	3 87½ by 64
	7		8
	0 15 2		31 00
	9		8
	Ans. 6 16 6		Ans. 248 00

	$	cts.			£	s.	d.			$	cts.
(7)	0	15¼×6	(8)		0	1	3	(9)		9	10×5
		10					12				10

	$	cts.			£	s.	d.			$	cts.
	1	52½			0	15	0			91	0×6
		10					11				10

	$	cts.		£	s.	d.		$	cts.
	15	25	Ans.	8	5	0		910	0
	0	91½							3

	$	cts.

	$	cts.
	2730	0
	546	0
	45	50

Ans. 16 16½

Ans. 3321 50

	£	s.	d.				$	cts.
(10)	0	9	6 per acre ×5		(11)		1	18¾×7
			10					10

	£	s.	d.		$	cts.
	4	15	0×2		11	87½×1
			10			10

	£	s.	d.		$	cts.
	47	10	0		118	75
			3			2

	£	s.	d.		$	cts.
	142	10	0		237	50
	9	10	0		11	87½
	2	7	6		8	31¼

Ans. 154 7 6

Ans. 257 68¾ prime cost.

Again : $1 $37\frac{1}{2} \times 7$
               10
_____

       13   75×1
               10
_____

     137   50
               2
_____

     275   00
      13   75
       9    $62\frac{1}{2}$
_____

  $298 $37\frac{1}{2}$ sold for.
  $257 $68\frac{3}{4}$ prime cost.
_____

  $40 $68\frac{3}{4}$ gain.
_____

COMPOUND SUBTRACTION.

EXAMPLES.

FEDERAL MONEY.

		$	cts.	m.
(2)	From	24	60	7
	Take	19	30	0
	Ans.	5	30	7

	$	cts.
(3)	600	$62\frac{1}{2}$
	1	75
Ans.	598	$87\frac{1}{2}$

	$	cts.
(4)	110	$18\frac{3}{4}$
	99	$10\frac{1}{4}$
Ans.	11	$8\frac{1}{2}$

	$	cts.	m.
(5)	960	10	2
		9	
Ans.	960	09	3

	$	cts.
(6)	449	$62\frac{1}{2}$
	1	$06\frac{3}{4}$
Ans.	448	$55\frac{3}{4}$

	$	cts.
(7)	1866	00
	278	$11\frac{3}{4}$
Ans.	1587	$88\frac{1}{4}$

	$	cts.			$	cts. m.			$	cts.
(8)	104	06¼		(9)	4010	14 4		(10)	400	00
		9¾			1011	12 5			211	12⅓
Ans.	103	96½		Ans.	2999	1 9		Ans.	188	87½

ENGLISH MONEY.

	£	s.	d.			£	s.	d.
(2)	47	6	7¾		(3)	419	7	6
	28	5	10½			227	8	9¼
Ans.	19	0	9¼		Ans	191	18	8¾

	£	s.	d.			£	s.	d.
(4)	1000	11	11¾		(5)	1000	2	4¼
	200	9	0			60	7	8¾
Ans.	800	2	11¾		Ans.	939	14	7¾

AVOIRDUPOIS WEIGHT.

	T.	cwt.	qr.	lb.	oz.	dr.			cwt.	qr	lb.	oz.
(2)	18	16	1	16	9	2		(3)	9	3	20	2
	0	19	3	20	0	6				2	23	5
Ans.	17	16	1	24	8	12		Ans.	9	0	24	13

	T.	cwt.	qr.	lb.			Cwt.	qr.	lb.
(4)	14	10	2	16		(5)	400	0	0
	0	0	0	11				2	3 14
Ans.	14	10	2	5		Ans.	397	0	14

TROY WEIGHT.

	lb.	oz.	dwt.	gr.			lb.	oz.	dwt.	gr.
(2)	8	3	0	2		(3)	106	0	0	15
	2	1	18	6			10	6	2	20
Ans.	6	1	1	20		Ans.	95	5	17	19

COMPOUND SUBTRACTION.

	lb.	oz.	dwt.	gr.			lb.	oz.	dwt.	gr.
(4)	22	0	12	6		(5)	16	0	0	0
	14	6	11	0			12	11	10	11
Ans.	7	6	1	6		Ans.	3	0	9	13

APOTHECARIES' WEIGHT.

	℔	℥	Ʒ	Ɖ	gr.			℔	℥	Ʒ			℔	℥	Ʒ
(2)	48	9	6	1	4	(3)	59	1	2		(4)	69	0	0
	1	10	0	2	8		,53	7	5			14	9	1
Ans.	46	11	5	1	16	Ans.	5	5	5		Ans.	54	2	7

CLOTH MEASURE.

	yd.	qr.	na.		yd.	qr.	na.		E.E.	qr.	na.		E.F.	qr.
(2)	950	1	2	(3)	49	0	2	(4)	66	4	0	(5)	44	1
	19	2	3		16	2	1		17	0	2		19	2
Ans.	930	2	3	Ans.	32	2	1	Ans.	49	3	2	Ans.	24	2

	E.Fl.	qr.				Yd.	qr.	na.		Yd.	qr.	na.
(6)	963	1	(7)	Bought		17	2	0	(8)	75	3	1
	174	2		Damaged		2	3	1		0	0	1
Ans.	788	2		Remains good		14	2	3	Ans.	75	3	0

LONG MEASURE.

	Deg.	m.	fur.	p.			M.	fur.	p.
(2)	20	50	4	20	(3) Travels first day		43	5	20
	11	56	0	30	second do.		32	4	00
Ans.	8	54	3	30	Ans.		11	1	20 more.

LAND MEASURE.

	A.	R.	P.			A.	R.	P.
(2)	502	2	10	(3)		69	1	3
	111	3	9			17	3	2
Ans.	390	3	1	Ans.		51	2	1

LIQUID MEASURE.

		T.	hhd.	gal.	qt.	pt.
(2)		100	1	19	2	1
		99	1	28	3	1
Ans.				3	53	3	0

		Hhd.	gal.
(3)		2	0
		0	29
Ans.	1	34

(4) From 1 pipe of wine, which is 126 gals., subtract 93 leaves 33 gals. of wine. Then from 4 hhds. of brandy subtract 29 gals., leaves 223 of brandy. Then from 2 bbls of beer, subtract 1, leaves 1 barrel, which is $31\frac{1}{2}$ gals.

Answer, 33 gals. wine, 223 gals. brandy, $31\frac{1}{2}$ gals. beer.

DRY MEASURE

	B.	p.	qt.	pt.			B.	p.	qt.	pt.			B.	p.	qt.	pt.
(2)	10	0	0	1	(3)	695	3	0	1	(5)	600	2	7	1
	9	2	6	1		589	3	5	0		146	3	2	1
Ans.		1	2	0	Ans.	105	3	3	1	Ans.	453	3	5	0

TIME.

	H.	min.	sec.
(2)	16	29	33
	7	36	44
Ans.	8	52	49

	Y.	m.	w.
(3)	18	11	2
	9	10	3
Ans.	9	0	3

	Y.	m.	w.	d.
(4)	900	0	0	0
	111	6	2	6
Ans.	788	6	1	1

	Y.	m.	w.	d.	h.
(5)	6	0	0	0	0
	1	1	1	1	1
Ans.	4	10	2	5	23

MOTION, OR CIRCLE MEASURE.

	sig.	°	′	″		sig.	°	′	″		sig.	°	′	′
(2)	9	7	40	8	(3)	10	10	16	12	(4)	11	2	5	14
	7	9	57	19		7	24	37	59		9	0	7	20
Ans.	1	27	42	49	Ans.	2	15	38	13	Ans.	2	1	52	54

(1) 6 feet of chain at $2,75
                per foot = $16 50
        A gold ring for   4 50
        Ear-rings   12 00

                        $33 00 whole amount.
                    Ring 4 50 has been returned.

                To receive $28 50

		$	cts.
(2)	2 doz. pairs at 75 cts.	= 18	00
	16 yds. at 87½ —	= 14	00
	28 do. at 22 —	= 6	16
	5 pair at 31¼ —	= 1	56¼

                Amount   39 72¼
                Note delivered 50 00

                Must be returned 10 27¾

		A.	R.	P.		£	s.	d.
(3)	1st tract contains	690	2	16	(4)	55	6	7
	2d do. do.	400	0	0		41	4	6
	3d do do.	63	3	24		75	0	0
	4th do. do.	63	3	24				

                                    Collected 171 11 1
        In the whole 1218 1 24 ·   Lost 40 6 0
        Sold   200 0 00
                                    I have 131 5 1
        Remains   1018 1 24

	Bu. p.		Bu. p.		Bu. p.	
(5) Bought	400 3	of wheat,	160 0	of rye,	150 2	of oats,
Sold	225 1	do.	37 2	do.	78 3	do.

Remaining 175 2   122 2   71 3

COMPOUND DIVISION.

EXAMPLES.

	$ cts.		$ cts.		$ cts.
(3)	3)366 18¾	(4)	6)384 87½	(5)	8)496 75

Ans. 122 6¼   Ans. 64 14½+2   Ans. 62 09¼+4

	$ cts.		$ cts.		$ cts.
(6)	9)587 68¾	(7)	11)976 43¾	(8)	12)1979 33⅓

Ans. 65 29¾+4   Ans. 88 76½+9   Ans. 164 94⅓+4

	£	s.	d.		£	s.	d.
(9)	3)560	9	7	(10)	5)475	19 ·	9¾

Ans. 186 16 6¼+1   Ans. 95 3 11½+1

	£	s.	d.		£	s.	d.
(11)	8)596	15	6½	(12)	12)756	4	1⅓¾

Ans. 74 11 11¼+2   Ans. 63 0 4¾+11

	Cwt.	qr.	lb.		Cwt.	qr.	lb.		Yds.	qr.	na.
(13)	5)45	3	27	(14)	9)10	0	15	(13)	7)44	1	2

Ans. 9 0 22+1   Ans. 1 0 14+1   Ans. 6 1 1+3

	Yds.	qr.	na.		M.	fur.	p.		M.	fur.	p.
(16)	11)56	3	3	(17)	12)105	5	22	(18)	6)45	7	18

Ans. 5 0 2+9   Ans. 8 6 18+6   Ans. 7 5 9+4

When the divisor exceeds 12, but is the exact product of any two figures in the multiplication table.

	$ cts. m.		$ cts. m.
(19)	6)45 66 5	(20)	4)98 77 8
	6)7 61 0+5	Rem.	11)24 69 4+2 Rem.

Ans. 1 26 8+2×6+5=17   Ans. 2 24 4+10×4+2=42

$cts. m.$ $ cts.$

21)12)77 87 5   (22) 12)283 68¾

     8)6 48 9+7   9)24 05½+1

            Rem.   Rem.

Ans.   0 81 1+1×12+7=19 Ans. 2 67¼+1×12+1=13

         $ cts. m.$

(23)   12)496 37 5

         11)41 36 4+7

      Ans.   3 76 0+4×12+7=55 Rem.

       £   s.   d.

(24)   4)87 19 4⅓

         8)21 19 10+2

     Ans.   2 14 11+6×4+2=²⁶⁄₈ qrs. =¾+2 Rem.

    £   s.   d.   £   s.   d.

25)3)55 4 7¾   (26) 8)97 15 6¼

   7)18 8 2½+1   7)12 4 5¼+1

           Rem.   Rem.

Ans.   2 12 7+6×3+1=19 Ans. 1 14 11+1×8+1=9

   $Hhd.gal.qt.$ $Hhd.gal.qt.$

27)7)44 28 2   (28) 12)150 47 3

   9)6 22 0+2   10)12 35 1+11

           Rem.   Rem.

Ans. 0 44 1+7×7+2=51 Ans. 1 16 0+5×12+11=71

When the divisor exceeds 12, and is not the product of any two figures in the multiplication table.

$ cts. $ cts. m.

(31) 78)196 75(2 52 2 Ans.
156

78)4075(52 cts.
3900

175
156

78)190(2 mills.
156

Rem. 34

$ cts. $ cts. m.

(32) 97)496 87½(5 12 2
485

97)1187(12 cts.
97

217
194

23
10

97)235(2 mills.
194

41 Rem.

$ cts. $ cts.

(38) 123)376 81¼(3 06¼ Ans
369

123)781(6 cts.
738

43
4

123)172(1
123

49 Rem.

£ s. d. £ s. d.

(34) 87)44 7 6(0 10 2¼ Ans.
20

87)887(10 shillings.
87

17
12

87)210(2 pence.
174

36
4

87)144(1 farthing.
87

57 Rem.

$$£ \quad s. \quad d. \quad £ \quad s. \quad d.$$

(35)   148)156   15   8¾(1   1   2¼ nearly, Ans.

            148

            ————

            8

           20

           ————

      148)175(1 shilling.

           148

           ————

           27

           12

           ————

     148)332(2 pence.

         296

         ————

         36

         4

         ————

         147

         148

         ————

PRACTICAL EXAMPLES.

$$\$ \quad cts. \quad m. \qquad\qquad\qquad \$ \quad cts. \quad \$ \quad cts.$$

(1)   6)47   87   5   (2) 112)64   81½(0   57¾ Ans.

     —————————   100

     4)7   97   9+1   —————

     —————————   112)6481(57 cts.

Ans. 1   99   4+3×6+1=19 Rem.   560

                       —————

                       881

                       784

                       —————

                       97

                       4

                       —————

                112)389(3

                      336

                      —————

                      53 Rem.

$ cts. $ cts. m.

(3) 72)56 25(0 78 1 Ans.
100

72)5625(78 cts.
504

585
576

9
10

72)90(1 mill.
72

18 Rem.

$ cts. $ cts. m.

(4) 63)125 00(1 98 4 Ans.
63

63)6200(98 cts.
567

530
504

26
10

63)260(4 mills.
252

8 Rem.

£ s. d.

(5) 4)18 17 6

Ans. 4 14 4½

$ cts. $ cts.

(6) 125)1875 81¼(15 00½ Ans.
125

625
625

125)00081(00 cts.
4

125)325(2 qrs.
250

75 Rem.

£ s. d. £ s. d.

(7) 1000)576 18 9¼(0 11 4¼ Ans.
20

1000)11358(11 shillings.
1000

1358
1000

358
12

1000)4305(4 pence.
4000

305
4

1000)1222(1 farthing.
1000

222 Rem.

```
        Gal. qt pt.G.qt.pt.              C. qr.lb.C.qr.lb.
(8)  89)150 2 1(1 2 1 Ans.       (9)  19)9 1 25(0 1 27 Ans.
        89                               4

        61                           19)37(1 qr.
         4                               19

     89)246(2 quarts.                    18
        178                              28

         68                             149
          2                              38

     89)137(1 pint.                  19)529(27 lbs.
        89                               38

        48 Rem.                         149
                                        133

                                         16 Rem.
```

—⋅●⋅—

REDUCTION.

FEDERAL MONEY.

EXAMPLES.

	$		$		$		cts.
(1)	10	(2)	25	(3)	387	(4)	25
	100		100		100		4
Ans.	1000	Ans.	2500	Ans.	38700	Ans.	100 fourths.

(5)　cts.
　　50
　　2
　———
Ans. 100 halves.

(6)　cts.
　　150
　　3
　———
Ans. 450 thirds.

(7)　$
　　50
　　100
　———
　5000
　　2
　———
Ans. 10000 halves.

(8)　$
　　25
　　100
　———
　2500
　　3
　———
Ans. 7500 thirds.

(9)　$
　　275
　　100
　———
　27500
　　4
　———
Ans. 110000 qrs.

(10)　$
　　10
　　10
　———
Ans. 100 dimes.

(11)　$
　　220
　　10
　———
　2200 dimes.
　　10
　———
　22000 cts.
　　10
　———
Ans. 220000 mills.

Note.—When more than one denomination is given to be reduced.

(1)　$ cts.
　　15 15
　　100
　———
Ans. 1515 cts.

(2)　$ cts.
　　2 25
　　100
　———
　225 cts.
　　4
　———
Ans. 900 4ths.

(3)　$ cts.
　　17 18¾
　　100
　———
　1718 cts.
　　4
　———
Ans. 6875 4ths.

E

	$	cts.			$	cts.
(4)	13	27⅓		(5)	426	88½
		100				100

1327		42688	
3		2	

Ans. 3982 thirds.   Ans. 85377 halves.

ENGLISH MONEY.

	£		s.		s.		d.
(2)	364	(3)	20	(4)	70	(5)	12
	20		12		12		4

Ans. 7280 s.   Ans. 240 d.   Ans. 840 d.   Ans. 48 qrs.

	d.		£ s. d.		£ s. d.		£ s. d.
(6)	26	(8)	18 12 7	(9)	105 13 9½	(10)	36 19 7¾
	4		20		20		20

Ans. 104 qrs.   372   2113   739
               12   12   12

       Ans. 4471 d.   25365   8875
                      4   4

            Ans. 101462   Ans. 35503 qrs.

Cents to Pence.

	cts.		cts.
(2)	36975	(3)	57697
	9		9

10)332775   10)519273

Ans. 33277½ d.   Ans. 51927¼+d.

Pence to Cents.

d.	*d.*
(2) 4590	(3) 76975
10	10
9)45900	9)769750
Ans. 5100 *cts.*	Ans. 85527 *cts.* 7 *m.*+7

AVOIRDUPOIS WEIGHT.

Cwt.	*qr.*	*lb.*	*oz.*
(2) 260	(3) 36	(4) 17	(5) 20
4	28	16	16
Ans. 1040 *qrs.*	288	102	120
	72	17	20
	Ans. 1008 *lbs.*	Ans. 272 *oz.*	Ans. 320 *dr.*

T. cwt. qr.	*Qr. lb. oz.*
(6) 5 12 2	(7) 2 25 10
20	28
112	21
4	6
Ans. 450 *qrs.*	81 lbs.
	16
	486
	82
	1306 ounces.
	16
	7836
	1306
	Ans. 20896 drams.

APOTHECARIES' WEIGHT.

	℥		℔		℔	℥	ℨ	Ə	gr.
(2)	72	(3)	10	(4)	15	9	4	2	17
	8		12		12				

Ans. 576 drams.

(3)
120 ozs.
8

960 drs.
3

2880 scru.
20

Ans. 57600 grs.

(4)
189 oz.
8

1516 drs.
3

4550 scru.
20

Ans. 91017 grs.

CLOTH MEASURE.

	Yds.		E. E.		E. Fl.
(2)	36	(3)	20	(4)	16
	4		5		3

Ans. 144 qrs.

Ans. 100 qrs.

48 qrs.
4

Ans. 192 na.

	E. Fl. qrs.		E. Fr. qr.		Yds. qrs. na.
(5)	5 2	(6)	37 2	(7)	19 2 1
	3		5		4

Ans. 17 qrs.

Ans 187 qrs.

78
4

Ans. 313 na.

DRY MEASURE.

	Pe.		Bu.		Bu.
(2)	32	(3)	7	(4)	12
	8		4		4
Ans.	256 *qts.*	Ans.	28 *pe.*		48
					8
					384
					2
				Ans.	768 *pts.*

	Bu.	pe.	qt.		Bu.	pe.	qt.	pt.
(5)	14	0	3	(6)	24	1	2	1
	4				4			
	56				97			
	8				8			
Ans.	451 *qts.*				778			
					2			
				Ans.	1557 *pts.*			

LAND MEASURE.

	A.		A.	R.	P.
(2)	132	(3)	54	3	23
	4		4		
	528		219		
	40		40		
Ans.	21120 *p.*	Ans.	8783		

SQUARE MEASURE.

	Sq. yds.			Sq.yds.	s.ft.	s.in
(2)	120		(3)	29	2	102
	9				9	

1080			263	
144			144	

4320			1054	
4320			1052	
1080			264	

Ans. 155520 *sq. in.*   Ans. 37974 *sq. in.*

LIQUID MEASURE.

	Gals.		Hhds		Gals.		Tuns.
(2)	28	(3)	5	(4)	110	(5)	6
	4		63		4		4

Ans. 112 *qts.*   Ans. 315 *gals.*

440	24
2	63

Ans. 880 *pts.*

72
144

	Hhds.	gals.	qts.		Gals.	qts.
(6)	7	41	2	(7)	47	2
	63				4	

1512				
4				

22			190	
46			2	

6048				
2				

482				
4				

Ans. 380 *pts.*   Ans. 12096 *pts*

Ans.1930 *qts.*

	Hhs.	gals.	qts.		Tuns	hhds.	gals.		Tun.	hhd.	gal.	qt.	pt.
(8)	4	0	3	(9)	19	0	27	(10)	5	1	15	1	1
	63				4				4				

(8)
63
252
4
1011
2
Ans. 2022 pts.

(9)
4
76 hhds.
63
235
458
4815
4
Ans. 19260 qts.

(10)
4
21
63
68
127
1338
4
5353
2
Ans. 10707 pts.

LONG MEASURE.

	Yds.		Po.		Fur.		Miles.
(2)	48	(3)	27	(4)	18	(5)	34
	3		5½		40		8

Ans. 144 ft.

(3)
135
13½
Ans. 148½ yds.

(4) Ans. 720 po.

(5) Ans. 272 fur.

	L.		M.		M.
(6)	108	(7)	17	(8)	20
	3		320 p.=1 m.		1760 yds.=1 m.

Ans. 324 m.

(7)
340
51
Ans. 5440 po.

(8) Ans. 35200 yds.

	ft.	in.			*Yds.*	*ft.*
		9	(11)		37	1
					3	

Ans. 112 *ft.*

Fur. po.
29

ft. in.
2 10

TROY WEIGHT

oz.
(2) 116
   20

Ans. 2320 dwt.

(3)
   lb.
   25
   12

  300
   20

 6000
   24

 24000
 12000

Ans. 144000 gr.

oz. dwt.
(4) 29 16
   20

Ans. 596 dwt.

lb.oz.dwt.gr.
(5) 19 11 14 21
   12

  239
   20

 4794
   24

 19177
  9590

Ans. 115077 gr.

TIME.

Min.
(1) 30
  60

Ans. 1800 s.

hrs.
(2) 12
  60

Ans. 720 m.

yrs.
(3) 12
  12

Ans. 144 m.

(4)
d. hr. min.
3 5 29
24

17
6

77
60

Ans. 4649 min.

	L.		*Ft. in.*		*Yds. ft.*
(9)	6	(10)	14 9	(11)	37 1
	3		12		3

	18	Ans. 177 *in.*	Ans. 112 *ft.*
	8		

	Fur. po.
144	(12) 112 29
40	40

5760	½)4509
5½	5½

28800	22545
2880	2254½

31680	Ans. 24799½ *yds.*
3	

	L. m. fur. po. yds. ft. in.
95040	(14) 2 1 3 16 3 2 10
12	3

Ans. 1140480 *in.*	7
	8

	M. fur. po.	59
(13) 450 6 32		40
8		

3606	½)2376
40	5½

Ans. 144272 *po.*	11883
	1188

	13071
	3

	39215
	12

Ans. 470590 *in.*

TROY WEIGHT

	oz.		lb.		oz. dwt.		lb.oz.dwt.gr.
(2)	116	(3)	25	(4)	29 16	(5)	19 11 14 21
	20		12		20		12

Ans. 2320 *dwt.* 300   Ans. 596 *dwt.*   239
20   20

6000   4794
24   24

24000   19177
12000   9590

Ans. 144000 *gr.*   Ans. 115077 *gr.*

TIME.

	Min.		hrs.		yrs.
(1)	30	(2)	12	(3)	12
	60		60		12

Ans. 1800 *s.*   Ans. 720 *m.*   Ans. 144 *m.*

	d. hr. min.
(4)	3 5 29
	24

17
6

77
60

Ans. 4649 *min.*

MOTION, OR CIRCLE MEASURE.

	°		Sig.		Sig. °		Sig. ° ′ ″
(1)	24	(2)	4	(3)	11 12	(4)	4 3 18 27
	60		30		30		30

Ans. 1440 ′   120   Ans. 342   123
                60   60

              7200   7398
                60   60

      Ans. 432000   Ans. 443907 ″

PROMISCUOUS EXAMPLES.

	$		Fur.		Days.		H. cts.
(1)	35	(2)	8)98	(3)	7)365	(4)	2)84
	100						

                  Ans. 12 m. 2 fur. Ans. 52 w. 1 d. Ans. 42 cts.
Ans. 3500 cts.

	Tuns cwt.		R.		S.		P.
(5)	8 15	(6)	63	(7)	2|0)15|7	(8)	4)175
	20		4⁰				

                         Ans. £7 17 s.   Ans. 43 bu. 3 pe.
Ans. 175 cwt. Ans. 2520 sq. per.

	cts.		Pts.		Sec.		Hhd. gal.
(9)	100)76|42	(10)	2)103	(11)	6|0)72|0	(12)	7 33
							63

    Ans. $76 42 cts. Ans. 51 qts. 1 pt. Ans. 12 min.
                                              24
                                              45

                                          Ans. 474 gal.

	Qrs.		*Dwts.*		*S.*

(13) 5)100   (14) 2|0)10|8   (15) 2|0)25|0

Ans. 20 *E. E.*   Ans. 5 *oz.* 8 *dwt.*   Ans. £12 10*s.*

(16) 3
7
3
———
Ans. 21 Ə

(17) *s. d.*
8 8
12
———
Ans. 104 *d.*

(18) *Days.*
7)203
———
Ans. 29 *w.*
———

(19) *Qrs.*
16
4
——
Ans. 64 *nu.*
——

(20) *drs.*
16)74(4*oz.*10 *drs.* Ans.
64
——
10
——

(21) *S.*
13
4
——
Ans. *threepences* 52
——

(22) *Tuns.*
20
20
——
Ans. 400 *cwt*
——

(23) *Qrs.*
5)81
——
Ans. 16 *E. Fr.* 1 *qr.*
——

(24) *Gal. qt. pt.*
21 3 1
4
——
87
2
——
Ans. 175 *pts.*
——

(25) *M. fur.*
3 1
8
——
Ans. 25 *fur.*
——

(26) *Cts.*
1|00)12|35
——
Ans. $12 35 *cts.*

(27) *Days.*
3
24
——
72
60
——
Ans. 4320 *m.*
——

(28) *Cts.*
121
4
———
Ans. 484 *qrs*
——

	lbs.		*Qrs.*		*Dwts.*

(29)   13   (30) 3)154   (31) 2|0)246|1
       16

Ans. 51 *E. Fl.* 1 *qr.* 12)123+1 *dwt.*

      —
      '78
      13   Ans. 10 *lb.* 3 *oz.* 1 *dwt*

     208
      16

   1248
    208

Ans. 3328 *drs.*

	Yd.	*qr.*	*na.*		*Gals.*

32)   12   2   1   (33) 63)584621(4)9279
     4   567.

     —   ——Ans. 2319 *t.* 3 *hhds.* 44*g.*
    50   176
     4   126

Ans. 201 *na.*   502
               441

	lbs.	*oz.*		611

(34)   725 6   567
      16

                44 *gals.*
  4356
   725   *lbs.*   *qrs.*
         (35) 28)27552(4)984
 11606   252
    16   —   246 *cwt.* Ans
            235
 69636   224
 11606

           112
Ans. 185696 *drs.*   112

(36)

£	s.	d.
5	4	6¼

20
—
104
12
—
1254
4
—

Ans. 5017 *fur.*

(37)

Days.

7)763
—
Ans. 109 *w.*
—

(38)

£	s.	d.
85	10	7

20
—
1710
12
—
Ans. 20527 *d.* ●

(39)

Grs.

2|0)122|0
—
3)61
—
Ans. 20 3 1 ϑ
—

(40)

Qrs.

5)27
—
Ans. 5 *E. E.* 2 *qrs.*
—

(41)

Pts.

2)1357
—
8)678 + 1 *pt.*
—
4)84 + 6 *qts.*
—
Ans. 21 *bu.* 0 *p.* 6 *qts.* 1 *pt.*
—

(42)

per.

4|0)865|4
—
4)216 + 14 *per.*
—
Ans. 54 *a.* 0 *r.* 14 *p.*
—

●━━●◉●━━●

SINGLE RULE OF THREE.

EXAMPLES.

lbs. lbs. cts. cts.

(3) State the question thus: As 2 : 8 :: 50 : 200 Ans.
For 50 × 8 = 400 ÷ 2 = 200 *cts.*

℔. lbs. cts. cts.

1) As 1 : 5 : : 12 : 60
   For 12 × 5 = 60 ÷ 1 = 60 *cts.* Ans.

yds. yd.   cts.   cts.

5) As 10 : 1 : : 550 : 55
   For 550 × 1 = 550 which ÷ 10 = 55 *cts.* Ans.

· lbs. lbs.   cts. $ cts.

6) As 7 : 122 : : 87½ : 15 25
   For 87½ × 122 = 10675 which ÷ 7 = $15 25 *cts.* Ans.

bu. bu.   cts. $ cts.

7) As 1 : 209 : : 72 : 150 48
   For 72 × 209 = 15048 which ÷ 1 = $150 48 *cts.* Ans.

lbs. lb.   cts.   cts.

8) As 5 : 1 : : 55 : 11
   For 55 × 1 = 55 which ÷ 5 = 11 *cts.* Ans.

yd. yds. $ cts. $ cts.

(9) As 1 : 18 : : 4 25 : 76 50
    For 425 × 18 = 7650 which ÷ 1 = $76 50 *cts.* Ans.

lbs. lb. $ cts.   cts.

(10) As 76 : 1 : : 24 32 : 32
     For 2432 × 1 = 2432 which ÷ 76 = 32 *cts.* Ans.

bu. bu. $ cts.   cts. m.

(11) As 8 : 1 : : 3 94 : 49 2 + 4
     For 394 × 1 = 394 which ÷ 8 = 49 *cts.* 2 *m.* + 4 Ans.

lb. lbs.   cts. $ cts.

(12) As 1 : 57 : : 7½ : 4 27½
     For 7½ × 57 = 427½ which ÷ 1 = $4 27½ *cts.* Ans.

bu. bu.   cts. $ cts.

(13)   As 1 : 243 : : 45 : 109 35
       For 45 × 243 = 10935 which ÷ 1 = $109 35 *cts.* Ans.

lb. lbs. $ cts. $ cts.

(14) As 1 : 147 : : 1 12½ : 165 37½   Ans.
     For 112½ × 147 = 16537½ which ÷ 1 = $165 37½ *cts.·*

℔. lbs.   cts. $ cts.

(15) As 1 : 869 : : 4½ : 39 10½
     For 4½ × 869 = 3910½ which ÷ 1 = $39 10½ *cts.* Ans.

    yds. *yd.* $ *cts.* $ *cts.*

(16) As 24 : 1 :: 125 24 : 5 21+20   Ans.

For 12524×1=12524 which ÷24=$5 21 *cts.*+20

    C. *lb.* $ *cts.* *cts. m.*

(17) As 1 : 1 :: 11 50 : 10 2+76

         lbs. *lb.* $ *cts.* *cts. m.*

Or, *e*3 112 : 1 :: 11 50 : 10 2+76   Ans.

For 1150×1=1150 which ÷112=10 *cts.* 2 *m.*+76

    lb. *lbs.* *cts.* $ *cts.*

(18) As 1 : 218 :: 7 : 15 26

For 7×218=1526 which ÷1=$15 26 *cts.* Ans.

    bu. *bu.* £ *s.* *s. d.*

(19) As 57 : 1 :: 30 10 : 10 8¼+39

       bu. *bu.* *s.* *s. d.*

Or, as 57 : 1 :: 610 : 10 8¼+

For 610×1=610 which ÷57=10*s.* 8¼*d.*+39 Ans.

    oz. *lbs.oz.* *cts.* $ *cts.*

(20) As 1 : 3 5 :: 72 : 29 52

      oz. *oz.* *cts.* $ *cts.*

Or, as 1 : 41 :: 72 : 29 52

For 72×41=2952 which ÷1=$29 52 *cts.* Ans.

    lb. *lbs.* *cts.* $ *cts.*

(21) As 1 : 135 :: 10 : 13 50

For 10×135=1350 which ÷1=$13 50 *cts.* Ans.

    C. *T. C.* £ *s. d.* £ *s. d.*

(22) As 2 : 15 3 :: 7 12 6 : 1155 3 9

      C. *C.*   *d.* £ *s. d.*

Or, as 2 : 303 :: 1830 : 1155 3 9

For 1830×303=554490 which ÷2=277245*d.*=
£1155 3*s.* 9*d.* Ans.

    s. d. £ *s. d.* *gal. gals.*

(23) As 4 10 : 54 7 6 :: 1 : 225

     d.   *d.*   *gal. gals.*

Or, as 58 : 13050 :: 1 : 225

For 1×13050=13050 which ÷58=225 *gals.* Ans.

    w. *w.* *cts.* $ *cts.*

(24) As 1 : 52 :: 250 : 130 00

For 250×52=13000 which ÷1=$130 00 *cts.* Ans.

A. A. R. P. $ *cts.* $ *cts.*

25) As 1 : 34 1 17 :: 42 25 : 1451 55+25

     P.   *P.* $ *cts.* $ *cts.*

Or, as 160 : 5497 :: 42 25 : 1451 55+25

For 4225×5497=23224825 which ÷160=$1451
55 *cts.*+25 Ans.

    gals. gal.   £ *s.*   *s.*

26) As 131 : 1 :: 65 10 : 10

     gals. gal.   *s.*   *s.*

Or, as 131 : 1 :: 1310 : 10

For 1310×1=1310 which ÷131=10*s.* Ans.

   $ $ *T. T. hhd. gal. qt. pt.*

(27) As 754 : 1754 :: 1 : 2 1 19 0 1

For 1×1754=1754 which ÷754=2 *T.* 1 *hhd.* 19
gal. 0 *qt.* 1 *pt.* Ans.

    s. d.   £ *s.*   *yds. yds.*

(28) As 18 8 : 36 16 :: 7 : 276

     d.   *d.*   *yds. yds.*

Or, as 224 : 8832 :: 7 : 276

For 8832×7=61824 which ÷224=276 *yds.* Ans.

   lb. cwt. qrs. lbs.   *cts.* $ *cts. m.*

(29) As 1 : 5 2 17 :: 9½ : 60 13 5

     lb.   *lbs.*   *cts.* $ *cts. m.*

Or, as 1 : 633 :: 9½ : 60 13 5

For 9½×633=6013½ which ÷1=$60 13 *cts.* 5 *m.* Ans.

   cts. $ *lb.*   *lbs. oz. dr.*

(30) As 114 : 354 :: 1 : 310 8 6+84

For 1×35400=35400 which ÷114=310 *lbs.* 8 *oz.*
6 *dr.*+84 Ans.

    £ *s.*   £ *s.*   *skeins. skeins.*

(31) As 2 10 : 105 3 :: 100 : 4206

     s.   *s.*   *skeins. skeins.*

Or, as 50 : 2103 :: 100 : 4206

For 100×2103=210300 which ÷50=4206 *sk.* Ans.

   yds. yd. $ *cts.* $ *cts. m.*

(32) As 39 : 1 :: 350 38 : 8 98 4+   Ans.

For 35038×1=35038 which ÷39=$8 98 *cts.* 4 *m.*

$$\text{gals. qts. gals. qt. pt. gals. qts. pt.}$$
(33) $61\tfrac{1}{2}$ gals.$=61$ $2+62$ 1 $1=123$ 3 1

$$pt. \quad gals. \; qts. pt. \quad cts. \quad \$ \; cts.$$
Then as 1 : 123 3 1 :: $37\tfrac{1}{2}$: 371 $62\tfrac{1}{2}$

$$pt \quad pts. \quad cts. \quad \$ \; cts.$$
Or, as 1 : 991 :: $37\tfrac{1}{2}$: 371 $62\tfrac{1}{2}$

For $37\tfrac{1}{2}\times 991=37162\tfrac{1}{2}$ which $\div 1=\$371\,62\tfrac{1}{2}$ cts.Ans.

$$bu. \quad bu. \quad bu.$$
(34) $75+87=162$

$$bu. \quad bu. \quad cts. \quad \$ \; cts.$$
Then as 1 : 162 :: 52 : 84 24

For $52\times 162=8424$ which $\div 1=\$84\,24$ cts. Ans.

(35) 1 year equals 365 days.

$$day. \quad days. \quad cts. \quad \$ \; cts.$$
Then as 1 : 365 :: $187\tfrac{1}{2}$: 684 $37\tfrac{1}{2}$

For $187\tfrac{1}{2}\times 365=68437\tfrac{1}{2}$ which $\div 1=\$684\,37\tfrac{1}{2}$ cts. the sum he spends in a year; his income yearly is $\$1022-\$684\,37\tfrac{1}{2}$ cts.$=\$337\,62\tfrac{1}{2}$ cts. Ans.

$$cwt. \; cwt. qrs. \; lb. \quad \$ \; cts. \quad \$ \; cts.$$
(36) As 1 : 4 3 24 :: 2 10 : 10 $42\tfrac{1}{2}$

$$lbs. \quad lbs. \quad cts. \quad \$ \; cts.$$
Or, as 112° : 556 :: 210 : 10 $42\tfrac{1}{2}$ price of stove.

For $210\times 556=110760$ which $\div 112=\$10\,42\tfrac{1}{2}$ cts. price of stove.

Then 27 lbs. $\times 18\tfrac{3}{4}$ cts.$=\$5\,06\tfrac{1}{4}$ cts. amount of pipe, and 50 cts. $\times 2=\$1.00$ price of 2 elbows.

$+\$10$ $42\tfrac{1}{2}$ cts..price of stove.
$+\$$ 5 $06\tfrac{3}{4}$ cts. do.   pipe.
$+\$$ 1 00 cts. do.   elbows.

———

$\$16$ $48\tfrac{3}{4}$ Ans.

(37) 14 pair$\times 2=28$ single shutters, which $\times 8\tfrac{1}{2}=243$: whole number of sheets used.

$$sheet \; sheets. \quad cts. \quad \$ \; cts.$$
Then as 1 : 243 :. $11\tfrac{1}{2}$: 27 37

For $243\times 11\tfrac{1}{2}=2737$ which $\div 1=\$27\,37$ cts. Ans.

38) If 45 men eat 1 lb. per day each, they will altogether eat 45 lbs. in a day.

     lbs.   *lbs.*   *d.* *w. d.*

Then as .45 : 4500 :: 1 : 14 2

For 1×4500=4500 which ÷45=100 d.=14 *weeks*
2 *days.* Ans.

    A. R. *A. R.P.* *bu. pe.* *bu.* *pe. qts.pt.*

39) As 12 2 : 37 3 5 :: 443 3 : 1341 0 7 1

    P.   *P.*   *pe.* *bu.* *pe. qts. pt.*

Or, as 2000 : 6045 :: 1775 : 1341 0 7 1

For 1775×6045=10729875 which ÷2000=1341 *bu.*
0 *pe.* 7 *qts.* 1 *pt.* Ans.

                $ cts.

40) Amount paid for the sugar 204 00
              carriage 15 75
              storage 18 31¼
     and would gain 57 00

               $295 06¼ the sum the
whole must sell for.

    C. qrs. C. $ cts. $ cts. m.

Then as 27 2 : 1 :: 295 06¼ : 10 72 9+60

    qrs. *qrs.* *cts.* $ cts. m.

Or, as 110 : 4 :: 29506¼ : 10 72 9+60

For 29506¼×4=118025 which ÷110=$10 72 *cts.*
9 *m.*+60 Ans.

(41) To find how much per cent. he can pay.

    $ cts. $ cts. $ $

As 18284 40 : 9142 20 :: 100 : 50 per cent.

For 100×914220=91422000 which ÷1828440=
50 Ans. the first.

To find what the creditor is to receive.

    $ cts. $ cts. $ $

As 18284 40 : 9142 20 :: 472 : 236

For 472×914220=431511840 which ÷1828440=
$236 Ans.

INVERSE PROPORTION.

   m. *m.* *d.* *d.*

(42) As 12 : 6 :: 18 : 9

  For 18×6=108 which ÷12=9 *days.* Ans.

   m. *m.* *d.* *d. h.*

(43) As 18 : 12 :: 20 : 13 4

  For 20×12=240 which ÷18=13 *days* 4 *hours.* Ans.

   d. *d.* *m.* *m*

(44) As 4 : 24 :: 8 : 48

  For 8×24=192 which ÷8=48 *men.* Ans.

   m. *m.* *d.* *d.*

(45) As 48 : 12 :: 24 : 6

  For 24×12=288 which ÷48=6 *days.* Ans.

   h. *h.* *d.* *d. h.*

(46) As 15 : 11 :: 5 : 3 8

  For 5×11=55 which ÷15=3 *days* 10 *hours.* Ans.

  ft. in. ft. in. *ft.* *yds.ft. in*

(47) As 2 3 : 30 6 :: 48 : 216 2 8

   in. *in.* *ft.* *yds.ft. in.*

  Or, as 27 : 366 :: 48 : 216 2 8

  For 48×366=17568 which ÷27=650$\frac{18}{27}$ *ft.* =216

  yds. 2 *ft.* 8 *in.* Ans.

   d. *d.* *m.* *m.*

(48) As 50 : 100 :: 14 , 28

  For 14×100=1400 which ÷50=28 *men.* Ans.

PROMISCUOUS EXAMPLES.

  Cwt. Cwt. qrs. lbs. $ *cts.*

(49) As 1 : 18 3 19 :: 11 37$\frac{1}{2}$

  Cwt. qrs. lbs. *lbs.*

  For 18 3 19=2119 which ×1137$\frac{1}{2}$=2410362$\frac{1}{2}$ the

  diviser; which ÷1 *cwt.*, that is 112 *lbs.*=$215

  21$\frac{5}{56}$ *cts.* Ans.

   $ $ $

(50) 730—22=708

   yds. *yd.* $ $ *cts. m.*

  Then as 156 : 1 :: 708 : 4 53 8+   Ans.

  For 708×1=708 which ÷156=$4 53 *cts.* 8 *m.*+72

(51) To find the prime cost.
    C. C. qrs. lbs. $ *cts.* $ *cts. m.*
    1 : 19   2   17 : : 9 $31\frac{1}{4}$: 183 00 7+
         lbs.   *lbs.* $ *cts.* $ *cts. m.*
Or, as 112 : 2201 : : 9 $31\frac{1}{4}$: 183 00 7+
For $931\frac{1}{4} \times 2201 = 2049681\frac{1}{4}$ which $\div 112 = \$183$
    00 *cts.* 7 *m.*+ Ans.

To find the sum it sold for.
     lbs.   *lbs.* $ *cts.* $ *cts. m.*
As 112 : 2201 : : 10 65 : 209 29 1+
For $1065 \times 2201 = 2344065$ which $\div 112 = \$209$ 29
    cts. 1 *m.* Ans.

To find the gain. It sold for $209 29 *cts.* 1 *m.*—
    $183 00 *cts.* 7 *m.* = $26 28 *cts.* 4 *m.*

       yds. yd. $ *cts. cts. m.*
(52) As 47 : 1 : : 14 75 : 31 3+
For $1475 \times 1 = 1475$ which $\div 47 = 31$ *cts.* 3 *m.*+ Ans.

(53) 3 *qrs.* wide : $1\frac{1}{4}$ wide : : $3\frac{3}{4}$ long : $6\frac{1}{2}$ long.
For $3\frac{3}{4} = 15$ *qrs.* and $1\frac{1}{4} = 5$ *qrs.* therefore $15 \times 5 =$
    75 which $\div 3 = 25$ *qrs.* = the quantity of holland
    requisite for each suit, and this 25 *qrs.* $\times 354$
    suits or men $= 8850$ *qrs.* which $\div 4 = 2212\frac{1}{2}$ *yds.*
    Ans.

(54) First 25 *ft.* : 250 *ft.* : : 33 *ft.* 10 *in.* : 338 *ft.* 4 *in.*
For 33 $10 \times 12 = 406$ *in.* $\times 250 = 101500$ which $\div 25$
    $= 4060$ *in.* = 338 *ft.* 4 *in.* the length of the shadow
    of the tower. Then as the shadow is 18 *ft.* 6 *in.*
    longer than the width of the river, consequently
    338 *ft.* 4 *in.* —18 *ft.* 6 *in.* = 319 *ft.* 10 *in.* the width
    of the river. Ans.

(55) First, 24 *hrs.* : 1 *m.* : : 360 *deg.* : 17 *m.* 3 *fur.* 1st
    Ans.
For $360 \times 69\frac{1}{2} \times 1 = 25020$ and 24 *hrs.* $\times 60 = 1440$;
    therefore $25020 \div 1440 = 17$ *m.* 3 *fur.*
Again, 24 *hrs.* : 1 *m.* : : 360 *deg.* : 11 *m.* 4 *fur.* =
    the velocity of the earth in lat. 40 *deg.*
For $360 \times 46 = 16560 \div 1440 = 11$ *m.* 4 *fur.*
Then, 17 *m.* 3 *fur.* —11 *m.* 4 *fur.* = 5 *m.* 7 *fur.* 2d Ans.

DOUBLE RULE OF THREE.

EXAMPLES.

(2) Thus 3 m. : 8 m. }
12 d. : 24 d. } :: 32A. : 170A. 2R.26P.20yds.+

For 8×24×32=6144 the dividend.
And 3×12=36 the divisor.
Then 6144÷36=170 A. 2 R. 26 P. 20 yds.+ Ans.

(3) Thus 10ox. : 20ox. } :: 2A. : 6A.
18d. : 27d. }

For 20×27×2=1080 the dividend.
And 18×10=180 the divisor.
Then 1080÷180=6 A. Ans.

(4) Thus 9m. : 2m. } :: 36 lbs. : 48 lbs.
10d. : 5d. }

For 24×5×36=4320 the dividend.
And 9×10=90 the divisor.
Then 4320÷90=48 lbs. Ans.

(5) Thus $100 : $335 } :: $6 : $30 15 cts.
12m. : 18m. }

For 335×18×6=36180 the dividend.
And 100×12=1200 the divisor.
Then 36180÷1200=$30 15 cts. Ans.

(6) Thus 20m. : 46m. } :: $56 31$\frac{1}{4}$ cts. : $828 92 cts.
5d. : 32d. }

For 46×32×5631$\frac{1}{4}$=8289200 the dividend.
And 20×5=100 the divisor.
Then 8289200÷100=$828 92 cts. Ans.

(7) Thus 8m. : 12m. } :: 120 $pairs$. : 540 $pairs$.
30d. : 90d. }

For 12×90×120=129600 the dividend.
And 8×30=240 the divisor.
Then 129600÷240=540 $pairs$. Ans.

(8) Thus 12p. : 38p. } :: 37lbs. : 468 lbs. 10$\frac{2}{3}$ oz.
4d. : 16d. }

For 38×16×37=22496 the dividend.
And 12×4=48 the divisor.
Then 22496÷48=468 lbs. 10$\frac{2}{3}$ oz. Ans.

(9) Thus $8li.$: $12li.$ $\Big\}$: : $5\,pts.$: $13\,pts.+$
　　　　$4E.$: $7E.$
For $12\times7\times5=420$ the dividend.
And $8\times4=32$ the divisor.
Then $420\div32=13+$ Ans.

(10) Thus $7\frac{1}{2}yds.$: $24yds.\ 2qrs.$ $\Big\}$: : $\$17\ 37\frac{1}{2}\,cts.$: $\$132$
　　　　　$3qrs.$: $7qrs.$ $\Big\}$ 　　　　　　$43\,cts.+$
For $24yds.\ 2qrs.=98qrs.$ And $7\frac{1}{2}yds.=30qrs.$
Then $98\times7\times1737\frac{1}{2}=1191925$ the dividend.
And $30\times3=90$ the divisor.
Then $1191925\div90=\$132\ 43\,cts.+$ Ans.

(11) Thus $20h.$: $62h.$ $\Big\}$: : $12bu.$: $60bu.\ 3pe.\ 3qts.\ 1pt.$
　　　　$22d.$: $36d.$ 　　　　　　　　　　　$+86$
For $62\times36\times12=26784$ the dividend.
And $20\times22=440$ the divisor.
Then $26784\div440=60bu.\ 3pe.\ 3qts.\ 1pt.+376$ Ans.

(12) Thus $\$100$: $\$563$ $\Big\}$: : $\$6$: $\$152\ 01ct.$
　　　　$12m.$: $54m.$
For $563\times18\times6=182412$ the dividend.
And $100\times12=1200$ the divisor.
Then $182412\div1200=\$152\ 01ct.$ Ans.

(13) Thus $8h.$: $20h.$ $\Big\}$: : $6T.$: $36T.\ 8cwt.\ 2qrs.\ 8lbs.$
　　　　$7m.$: $17m.$
For $20\times17\times6=2040$ the dividend.
And $8\times7=56$ the divisor.
Then $2040\div56=36T.\ 8cwt.\ 2qrs.\ 8lbs.$ Ans.

(14) Thus $2yds.$: $50yds.$ $\Big\}$: : $1lb.$: $15lbs.$
　　　　$5qrs.$: $3qrs.$
For $50\times3\times1=150$ the dividend.
And $2\times5=10$ the divisor.
Then $150\div10=15lbs.$ Ans.

(15) Thus $\$21$: $\$96$ $\Big\}$: : $7re.$: $3re.$
　　　　$32d.$: $3d.$
For $96\times3\times7=2016$ the dividend.
And $21\times32=672$ the divisor.
Then $2016\div672=3.$ Ans.

(16) Thus $4m.$: $12m.$ $\Big\}$ :: $100 : $360
$7\frac{1}{2}$: $9

For $12\times9\times100=10800$ the dividend.
And $4\times7\frac{1}{2}=30$ the divisor.
Then $10800\div30=$360$. Ans.

(17) Inversely thus $40ft.$ $\Big\}$: $\Big\{$ $20ft.$ $\Big\}$
$54ft.$: $54ft.$ $\Big\}$ :: $10d.$: $1d.$ $10\frac{1}{2}hrs.$
$72m.$: $27m.$

For $20\times54\times27\times10=291600$ the dividend.
And $40\times54\times72=155520$ the divisor.
Then $291600\div155520=1d.$ $10\frac{1}{2}hrs.$ Ans.

(18) Thus $305m.$: $1056m.$ $\Big\}$ :: $30d.$: $116d.+2540$
$12\frac{1}{2}h.$: $14h.$

For $1056\times14\times30=443520$ the dividend.
And $305\times12\frac{1}{2}=3812\frac{1}{2}$ the divisor.
Then $443520\div3812\frac{1}{2}=116d.$ Ans.

(19) Thus $210 : $837 $\Big\}$ :: $24w.$ $3d.$: $25w.$ $6d.+$
$15m.$: $4m.$

For $24w.$ $3d.=171d.$ And $837\times4\times171=572508$
the dividend.
And $210\times15=3150$ the divisor.
Then $572508\div3150=181d.=25w.$ $6d.+2358$ Ans.

(20) Thus $2\frac{1}{2}yrs.$: $5yrs.$ $\Big\}$ :: $50 : $200
$15 : $30

For $5\times30\times50=7500$ the dividend.
And $2\frac{1}{2}\times15=37\frac{1}{2}$ the divisor.
Then $7500\div37\frac{1}{2}=$200$. Ans.

(21) Thus $5m.$: $34m.$ $\Big\}$ :: $20 50 cts. : $3136 50 cts.
$4d.$: $90d.$

For $34\times90\times2050=6273000$ the dividend.
And $5\times4=20$ the divisor.
Then $6273000\div20=$3136 50 cts.$ Ans.

(22) Thus $24cwt.$: $76cwt.$ $\Big\}$ :: $18 : $153 26 cts.+720
$45m.$: $121m.$

For $76\times121\times18=165528$ the dividend.
And $24\times45=1080$ the divisor.
Then $165528\div1080=$153 26 cts.$ Ans.+

23) Thus $42A. : 392A.$ $\Big\}$ :: $6men : 28men.$
　　　　$14D. : 7D.$

For $392 \times 7 \times 6 = 16464$ the dividend.

And $42 \times 14 = 588$ the divisor.

Therefore $16464 \div 588 = 28men$. **Ans.**

24) Thus $35cwt. : 50cwt.$ $\Big\}$ :: $\$9\ 50cts. : \$101\ 78\frac{1}{2}cts.+$
　　　　$20m. : 150m.$

For $50 \times 150 \times 950 = 7125000$ the dividend.

And $35 \times 20 = 700$ the divisor.

Then $7125000 \div 700 = \$101\ 78\frac{1}{2}cts.+$ **Ans.**

25) Thus $\$11\ 75cts. : \$31\quad 18\frac{3}{4}cts.$ $\Big\}$ :: $\$125 : \$663\ 56_2$
　　　　$9m. : 1yr.\ 6mo.$ $\qquad\qquad cts.+$

For $3118\frac{3}{4} = 12475qrs. \times 18m. \times 125 = 28068750$ the
　　dividend.

And $\$11\ 75cts. = 4700qrs. \times 9 = 42300$ the divisor.

Then $28068750 \div 42300 = \$663\ 56\frac{1}{4}cts.+$ **Ans.**

26) Thus $\$100\quad : \$275$ $\Big\}$ :: $\$6 : \$77$
　　　　$12m. : 56m.$

For $275 \times 56 \times 6 = 92400$ the dividend.

And $100 \times 12 = 1200$ the divisor.

Then $92400 \div 1200 = \$77$. **Ans.**

(27) Thus $\$56\quad : \$6$ $\Big\}$ :: $\$560 : \$100$
　　　　$12m. : 20m.$

For $6 \times 20 \times 560 = 67200$ the dividend.

And $56 \times 12 = 672$ the divisor.

Then $67200 \div 672 = \$100$. **Ans.**

(28) Thus $12yds. : 75yds.$ $\Big\}$ :: $5lb. : 18lbs.\ 12oz.$
　　　　$5qrs. : 3qrs.$

For $75 \times 3 \times 5 = 1125$ the dividend.

And $12 \times 5 = 60$ the divisor.

Then $1125 \div 60 = 18lb.\ 12oz.$ **Ans.**

PRACTICE.

CASE 1.

(3) $|\frac{1}{2}|\frac{1}{2}|$296 at $\frac{3}{4}$

　　　$|\frac{1}{4}|\frac{1}{2}|$148
　　　　　74

Ans. $2 22 *cts.*

(4) $|\frac{1}{2}|\frac{1}{2}|$3268 at $\frac{1}{2}$

Ans. $16 34 *cts.*

(5) $|\frac{1}{2}|\frac{1}{2}|$4260 at $\frac{3}{4}$

　　　$|\frac{1}{4}|\frac{1}{2}|$2130
　　　　　1065

Ans. $31 95 *cts.*

(6) $|\frac{1}{4}|\frac{1}{4}|$5324 at $\frac{1}{4}$

Ans. $13 31 *cts.*

m.

(7) $|2|\frac{1}{4}|$634 at 2 *mills.*

Ans. $1 26 8

m.

(8) $|2|\frac{1}{3}|$352 at 4 *mills.*

　　　$|2|\frac{1}{5}|$ 70 4
　　　　　 70 4

Ans. $1 40 8

m.

(9) $|5|\frac{1}{2}|$3456 at 5 *mills.*

Ans. $17 28

m.

(10) $|5|\frac{1}{2}|$498 at 6 *mills.*

　　　$|1|\frac{1}{5}|$249
　　　　　 49 8

Ans. $2 98 8

$m.$

(11) $|5|\frac{1}{2}|$8462 at 8 *mills*

$|2|\frac{1}{5}|$4231
$|1|\frac{1}{2}|$1692 4
846 2

Ans. $67 69 6

$m.$

(12) $|5|\frac{1}{2}|$1264 at 7 *mills.*

$|2|\frac{1}{5}|$632
252 8

Ans. $8 84 8

$m.$

(13) $|5|\frac{1}{2}|$4628 at 9 *mills.*

$|2|\frac{1}{5}|$2314
$|2|\frac{1}{5}|$925 6
925 6

Ans. $41 65 2

CASE 2

cts.

(2) $|6\frac{1}{4}|\frac{1}{16}|$3648 at 6¼ *cts.*

Ans. $228

cts.

(3) $|10|\frac{1}{10}|$742 at 10 *cts.*

Ans. $74 20

cts.

(4) $|20|\frac{1}{5}|$8264 at 20 *cts.*

Ans. $1652 80

cts.

(5) $|25|\frac{1}{4}|$386 at 25 *cts.*

Ans. $96 50

cts.

(6) $|50|\frac{1}{2}|$5876 at 50 *cts.*

Ans. $2938

cts.

(7) $|25|\frac{1}{4}|$3542 at 45 *cts.*

$|20|\frac{1}{5}|$885 50
708 40

Ans. $1593 90

(8)

cts.		
50	$\frac{1}{2}$	31925 at 80 *cts.*
25	$\frac{1}{2}$	15962 50
5	$\frac{1}{5}$	7981 25
		1596 25
	Ans.	$25540 00

(9)

cts.		
12½	$\frac{1}{8}$	4264 at 12½ *cts.*
	Ans.	$533

(10)

cts.		
50	$\frac{1}{2}$	18626 at 55 *cts.*
5	$\frac{1}{10}$	9313
		931 30
	Ans.	$10244 30

(11)

cts.		
25	$\frac{1}{4}$	1724 at 37½ *cts.*
12½	$\frac{1}{2}$	431
		215 50
	Ans.	$646 50

(12)

cts.		
10	$\frac{1}{10}$	528 at 16 *cts.*
5	$\frac{1}{2}$	52 80
1	$\frac{1}{5}$	26 40
		5 28
	Ans	$84 48

(13)

cts.		
50	$\frac{1}{2}$	13854 at 56¼ *cts.*
6¼	$\frac{1}{8}$	6927
		865 87 5
	Ans.	$7792 87 5

(14)

cts.		
20	$\frac{1}{5}$	4858 at 29 *cts*
5	$\frac{1}{4}$	971 60
4	$\frac{1}{5}$	242 90
		194 32
	Ans.	$1408 82

(15)

cts.		
50	$\frac{1}{2}$	2267 at 85 *cts.*
25	$\frac{1}{2}$	1133 50
10	$\frac{1}{5}$	566 75
		226 70
	Ans.	$1926 95

(16)

cts.		
20	$\frac{1}{5}$	190 at 20 *cts.*
	Ans.	$38

(17)

cts.				
$12\frac{1}{2}$	$\frac{1}{8}$	3654 at $18\frac{3}{4}$ *cts.*		
$6\frac{1}{4}$	$\frac{1}{2}$	456	75	
		228	37	5
Ans.		$685	12	5

(18)

cts.			
50	$\frac{1}{2}$	17638 at 70 *cts.*	
10	$\frac{1}{5}$	8819	
10	$\frac{1}{5}$	1763	80
		1763	80
Ans.		$12346	60

CASE 3.

(2)

		$	cts.	
2	$\frac{1}{2}$	10	25	
			10	
		102	50	
7	$\frac{1}{8}$	5	12	5
		0	64	0
Ans.		$108	26	5

(3)

		$	cts.	
2	$\frac{1}{2}$	4	15	
			7	
		29	05	
1	$\frac{1}{2}$	2	07	5
14	$\frac{1}{2}$	1	03	7
4	$\frac{1}{7}$	0	51	8
1	$\frac{1}{4}$	0	14	8
		0	3	7
Ans.		$32	86	5

Cwt. qr. lb.　$ cts.
(4)　129　1　10　at　1　05
　　　　　　　　　129

　　　　　　　945
　　　　　　　210
　　　　　　　105

1 |　½ | 13545
7 |　¼ |　　26　2
2 | 1/14 |　　6　5
1 |　⅛ |　　1　8
　　　　　　0　9

　Ans. $135　80　4

Cwt. qr. | $
(6)　130 1 at 15
　　　　　130

1 | 1¼ | 450
　　　　 15

　　　 | 1950
　　　　　3　75

　Ans. $1953　75

qrs. lb.　　cts.
(8)　2　14　at　2710

2 | 1/3 | 1355
14 | 1/4 | 338. 7

　Ans. $16　93　7

Cwt. qr.　$ cts.
(5)　16　2　at　5　18
　　　　　　　　16

2 | 1½ | 3108
　　　　 518

　　 | 82　88
　　 | 2　59

　Ans. $85　47

Cwt. qr. lb.　cts.
(7)　25　1　9　at　175
　　　　　　　　　25

1 | 1½ | 875
4 | 1/7 | 350

4 | 1¼ | 43　75
1 | 1¼ | 43　7
　　　　　6　2+
　　　　　6　2+
　　　　　1　5+

　Ans. $44　32　8

lb. oz. dwt. gr.　$ cts.
(9)　6　5　10　5　at　4 16
　　　　　　　　　　　6

4 | 1½ |
1 | 1/2/7 | 2496
10 | 1/2/7 | 138　6
5 | 4 8 | 34　6
　　　　 17　3
　　　　　　3

　Ans. $26　86　8

	lb. oz. dwt. gr.	cts.			lb. oz. dwt.gr.	cts.

(10) 27 10 4 18 at 2635   (11) 9 11 17 22 at 613
              27   9

                                            6 |½|
       6 |½| 18445   4 |⅓| 5517
            5270   1 |¼| 306 5
                           10 |½| 204 3
       3 |½| 711 45   5 |⅓| 51 0
       1 |⅔| 13 17 5   2 |⅓| 25 5
       4 | 6 58 7   12 |¼| 12 7
     12 | 2 19 5   6 |½| 5 1
       6 | 43 9   2 |⅓| 1 2
               5 4   2 |⅓| 6
               2 7   2
                                        2

Ans. $733 92 7   Ans. $61 24 3

	oz. dwt. gr.	cts.			yd. qr.	$ cts.

(12) 816 13 12 at 12½   (13) 27 3 at 9 65
             816   27

       10 |½| 1632   2 |½| 6755
            816   1930
            408

           102 00   260 55
       2 |½| 6 2   1 |½| 4 82 5
       1 |⅔| 1 2   2 41 2
     12 |½| 6
               3   Ans. $267 78 7

Ans. $102 08 3

	yd.	qr.	cts.
(14)	860	1 at	84
			860

1	¼	5040
		672

		722	40
			21

Ans. $722 61

	yd.	qr.	na.	cts.
(15)	126	2	2 at	475
				126

2	½	2850
		950
		475

		598	50	
2	¼	2	37	5
		59	3

Ans. $601 46 8

	gal.	qt	cts
(16)	428	3 at	140
			428

2	½	1120
		280
		560

		599	20
1	½	70
		35

Ans. $600 25

	gal.	qt.	pt.	cts.
(17)	765	3	1 at	218¾
				4

2	½	875
		765

		4375
		5250
		6125

		6693	75
1	½	4	37
1	½	2	18
		1	09

4)6701 39

Ans. $1675 34¾

```
        hhd. gal.    $ cts.              hhd. gal. qt.   $ cts.
(18)     5    31½  at 47 12      (19)     17   15   3 at 64 75
                       5                                    17

        |31½|½|235 60                    |9|1/7|453 25
              |23 56                            |647 5

        Ans. $259 16                           |1100 75
                                          |3|1/3|   9 25
                                          |3|1/3|   3 08 3
        bu.  pe.    cts.                  |3|1/4|   3 08 3
(20)    120   2  at 35                             77 1

                    120                   Ans. $1116 93 7

        |2|½|700                         bu.  pe. qt.   $ cts.
             |35                 (21)    780   3   2 at 1 17
                                                        780
        |4200
         17 5                             |2|½|9360
                                               |819
        Ans. $42 17 5
                                              |912 60
                                          |1|1/2| 58 5
                                          |2|1/4| 29 2
        bu. pe.qt.pt.   cts.                       7 3
(22)   1354 1 5  1  at 25
                   1354          Ans. $913 55 0

        |1|¼|100                         A.  R.  P.    $ cts.
            |125                 (23)    35   2  18 at 54 35
            |75                                          35
            |25
                                          |2|½|27175
        |338 50                               |16305
        |4|½|  6 2½
        |1|¼|  3 1¼                          |1902 25
        |1|½|  7¾                        |16|1/2| 27 17 5
               3¾                        |2|1/8|  5 43 5
                                                  67 9
        Ans. $338 60 5¼
                                Ans. $1935 53 9
```

A.R.P. $ cts.
(24) 146 3 10 at 35 10
          146

2	$\frac{1}{2}$	21060
		14040
		3510

		5124 60
1	$\frac{1}{2}$	17 55
10	$\frac{1}{2}$ $\frac{1}{4}$	8 77 5
		2 19 3+

Ans. $5153 11 8+

A. R.P. $ cts.
(25) 750 1 4 at 12 25
          750

1	$\frac{1}{4}$	61250
		8575

		9187 50
4	$\frac{1}{16}$	3 06 2$\frac{1}{2}$
		0 30 6$\frac{1}{4}$

Ans. $9190 86 8$\frac{3}{4}$

APPLICATION.

Cwt. qr. lb. $ cts.
(1) 84 2 14 at 10 50
         84

2	$\frac{1}{2}$	4200
		8400

		882 00
14	$\frac{1}{4}$	5 25
		1 31 2+

Ans. $888 56 2+

Cwt. qr. lb. cts.
(2) 17 1 7 at 1212$\frac{1}{2}$
         2

1	$\frac{1}{4}$	2425 halves.
		17

		16975
		2425

		412 25
7	$\frac{1}{4}$	6 06 1$\frac{1}{2}$
		1 51 5$\frac{1}{2}$

2)419 83 8 mills.

Ans. $209 91 9 mills.

```
        T.cwt.qr.  $  cts.              yd. qr. pie.  yd.
(3)  15 10  3 at 80  15       (4)  35  2×170=6035 at ¼
                     15                          6035
                                        _____
      |10|½| 40075                          4)6035 qrs.
            8015
                                         Ans. $15 08
           _____
           1202 25
      |2|1/20| 40 07 5
      |1| ½ |  2 00 3¾
               1 00 1¾

       Ans. $1245 33 0½
```

```
                A. R. P.   $ cts.
        (5)   175 3 12 at 52 15
                            175
                      _____
             |2|½|   26075
                     36505
                     5215
                     _____
                      9126 25
             |1|1/5|   26 07 5
             |10|1/4|  13 03 7
             |2|1/3|    3 25 9
                        0 65 1

             Ans. $9169 27 2
```

```
(6)  1365 at ½ct.=$6 82½cts. Ans.     (7)  784 at 84 cts.
                                                     784
                                               _____
                                                     336
                                                     672
                                                     588
                                               _____
                                            Ans. $658 56
```

STERLING MONEY.
CASE 1.

(4) $\frac{1}{4}|\frac{1}{4}|$ 475 at $\frac{1}{4}$

12)118$\frac{3}{4}$

Ans. 9s. 10$\frac{3}{4}$d.

(5) $\frac{1}{2}|\frac{1}{2}|$ 299 at $\frac{1}{2}$

12)149$\frac{1}{2}$

Ans. £12s. 5$\frac{1}{2}$d.

(6) $\frac{1}{2}|\frac{1}{3}|$ 978 at $\frac{3}{4}$

$\frac{1}{4}$ $\frac{1}{2}$ 489

244$\frac{1}{2}$

12)733$\frac{1}{2}$

2|0)6|1 1

Ans. £3 1s. 1$\frac{1}{2}$d.

CASE 2.

(2) $2|\frac{1}{6}|$ 978 at 2d.

2|0)16|3

Ans. £8 3s.

(3) $4|\frac{1}{3}|$ 499 at 5d.

$1|\frac{1}{4}|$ 166 4

41 7

2|0)20|7 11

Ans. £10 7s. 11d.

(4) $6|\frac{1}{2}|$ 792 at 6d.

2|0)39|6

Ans. £19 16s.

(5) $6|\frac{1}{2}|$ 888 at 9d.

$3|\frac{1}{2}|$ 444

222

2|0)66|6

Ans. £33 6s.

(6) | 6 | $\frac{1}{2}$ | 921 at 11*d.*

| 3 | $\frac{1}{2}$ | 460 6
| 2 | $\frac{1}{3}$ | 230 3
153 6

2|0)84|4 3

Ans. £42 4*s.* 3*d.*

CASE 3.

(2) | 3 | $\frac{1}{4}$ | 487 at 15*d.*
121 9

2|0)60|8—9

Ans. £30 8*s.* 9*d.*

(3) | 6 | $\frac{1}{2}$ | 979 at 22$\frac{1}{4}$
| 3 | $\frac{1}{2}$ | 489 6
| 1 | $\frac{1}{3}$ | 244 9
| $\frac{1}{4}$ | $\frac{1}{4}$ | 81 7
20 4$\frac{3}{4}$

2|0)181|5 2$\frac{3}{4}$

Ans. £90 15*s.* 2$\frac{3}{4}$*d.*

(4) | 6 | $\frac{1}{2}$ | 532 at 23$\frac{3}{4}$*d.*
| 4 | $\frac{1}{3}$ | 266
| 1 | $\frac{1}{4}$ | 177 4
| $\frac{1}{2}$ | $\frac{1}{2}$ | 44 4
| $\frac{1}{4}$ | $\frac{1}{2}$ | 22 2 $\frac{1}{2}$
11 1 $\frac{1}{4}$

2|0)105|2 11$\frac{3}{4}$

Ans. £52 12*s.* 11$\frac{3}{4}$*d.*

CASE 4.

(2) | 5 | $\frac{1}{4}$ | 489 at 5*s.*

Ans. £122 5*s.*

(3)

10	$\frac{1}{2}$	937 at 11*s.*		
1	$\frac{1}{10}$	468	10	
		46	17	
		Ans. £515	7*s.*	

(4)

10	$\frac{1}{2}$	1286 at 15*s.*		
5	$\frac{1}{2}$	643		
		321	10	
		Ans. £964	10*s.*	

(5)

10	$\frac{1}{2}$	2798 at 19*s.*	
5	$\frac{1}{2}$	1399	
4	$\frac{1}{5}$	699	10
		559	12
		Ans. £2658	2*s.*

CASE 5.

(2

		£	*s.*	*d.*
10	$\frac{1}{2}$	569 at 4 13 $7\frac{1}{2}$		
		4		
		2276		
2	$\frac{1}{5}$	284	10	
1	$\frac{1}{2}$	56	18	
6	$\frac{1}{2}$	28	9	
1	$\frac{1}{6}$	14	4	6
$\frac{1}{2}$	$\frac{1}{2}$	2	7	5
		1	3	$8\frac{1}{2}$
		Ans. £2663	12	$7\frac{1}{2}$

(3)

		£	*s.*	*d.*
10	$\frac{1}{2}$	1967 at 5 16 $9\frac{3}{4}$		
		5		
		9835		
5	$\frac{1}{2}$	983	10	
1	$\frac{1}{5}$	491	15	
6	$\frac{1}{2}$	98	7	
3	$\frac{1}{2}$	49	3	6
$\frac{3}{4}$	$\frac{1}{4}$	24	11	9
		6	2	$11\frac{1}{4}$
		Ans. £11488	10	$2\frac{1}{4}$

(4) 2975 at £7 19*s*. 11¾*d*.

10	½	2975 at £7 19*s*. 11¾*d*.		
		7		
		20825		
5	½	1487	10	
4	⅕	743	15	
8	⅙	595		
2	¼	99	3	4
1	½	24	15	10
½	½	12	7	11
¼	½	6	3	11½
		3	1	11¾
	Ans.	£23796	18	0¼

CASE 6.

(2) *C. qr. lb.* 9 2 17 at £4 7*s*. 6*d*.

		£	*s.*	*d.*
		4	7	6
				9
2	½	39	7	6
14	¼	2	3	9
2	⅐	0	10	11¼
1	½		1	6¾
				9¼
	Ans.	£42	4	6¼

(3) *C. qr. lb.* 11 1 16 at £5 6*s*. 7½*d*.

		£	*s.*	*d.*
		5	6	7½
				11
1	¼	58	12	10½
14	½	1	6	7¾
2	⅐		13	3¾
			1	10¾
	Ans.	£60	14	8¾+

(4) *C. qr. lb.* 7 3 22 at £1 18*s*. 4¾*d*.

		£	*s.*	*d.*
		1	18	4¾
				7
2	½	13	8	9¼
1	½	0	19	2¼
14	½		9	7
7	½		4	9½
1	⅐		2	4¾
				4
	Ans.	£15	5	0¾

(5) *C. qr. lb.* 27 1 19 at £2 17*s*. 8¼*d*. 3×9=27

		£	*s.*	*d.*
1	¼	8	13	0¾
				9
		77	17	6¾
14	½		14	5+
4	⅐		7	2½
1	¼		2	0½
				6+
	Ans.	£79	1	8¾

TARE AND TRET.

CASE 1.

	Cwt.	qr.	lb.		Cwt.	qr.	lb.		Cwt.	qr.	lb.
(2)	7	3	20	(3)	6	2	5	(4)	369	2	21
			8		—	1	11		—10	1	12
gross	63	1	20	Ans.	6	0	22	Ans.	359	1	9
	—5	1	19								
Ans.	58	0	1								

	Cwt.	qr.	lb.
(5)	5	1	19
			8
	43	1	12	whole gross.
	—2	0	23	tare.
Ans.	41	0	17	neat.

(6)

	C.	qr.	lb.	lb.
No. 1.	3	2	19	tare 34
No. 2.	6	0	13	tare 57
No. 3.	4	3	5	tare 46

	C.	qr.	lb.
14 2 9 w. t. 137	= 1	0	25
—1	0	25
Ans. 13 1 12			

CASE 2.

	C.	qr.	lb.
(2)	4	2	24
			7
	33	0	0	gross.
	4	2	14	tare.
Ans.	28	1	14	neat.

qr.	lbs.
2	18
	7

4cwt. 2qrs. 14lbs. whole tare.

```
              C.  qr.  lb.
  (3)         21   2   21
               3   0   18
            _____        $ cts.
    Neat 18   2    3  at 5 50
                                      18
                        |qrs.|    |
                        |  2 | ½ |4400
                        |    |   | 550
                        |    |   |_____
                        |lbs.|   |9900
                        |  2 |1/20| 275
                        |  1 |  ½ |   9 8+
                        |    |   |   4 9·
                            _____
                    Ans. $101 89 7
```

```
              C.  qr.  lb.              lb
  (4)          2   1   25               30 ·
                       9                 9
            _____           ____     C.  qr.  lb.
            22   1    1 gross.         270=2  1  18
             2   1   18 tare.
            _____           $ cts.
    Neat 19   3   11  at 5 10
                                        19
                        |qrs.|    |
                        |  2 | ½ |45 90
                        |    |   |51  0
                        |    |   |_____
                        |    |   |96 90
                        |  1 |1/20| 2 55
                        |  7 |1/20| 1 27 5
                        |  4 |1/20|   31 8+
                        |    |   |   18 2+
                            _____
                    Ans. $101 22 5 value.
```

CASE 3.

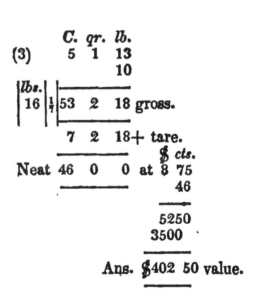

```
                C. qr. lb.
(2)             7   3   14
                        4
    |lbs.|  |
    | 16 |⅟₇|31   2    0 gross.

    |  4 |¼|  4   2    0
             1   0   14

              5   2   14 tare.

    Ans.  25   3   14 neat.
```

```
              C. qr. lb.
(3)           5   1   13
                     10
   |lbs.|  |
   | 16 |⅟₇|53  2   18 gross.

              7  2   18+ tare.
                          $ cts.
   Neat   46  0    0 at 8 75
                           46

                         5250
                         3500

          Ans.  $402 50 value.
```

(4)   4C. 1qr. 24lb.
$$6$$

$\frac{lbs.}{16}$ $\frac{1}{7}$	26	3	4 gross.
2 $\frac{1}{8}$	3	3	8+
	0	1	25+

Tare   4   1   5

Neat 22   1   27 = 2519 at $7\frac{1}{2}$
$$7\frac{1}{2}$$
17633
1259 5

Ans. $188 92 5 value.

CASE 4.

(2)   2C. 1qr. 10lb.
$$12$$

$\frac{lbs.}{16}$ $\frac{1}{7}$	28	0	8 gross.
2 $\frac{1}{8}$	4	0	1
	0	2	0
	4	2	1 tare.
$\frac{lbs.}{4}$ $\frac{1}{26}$	23	2	7 suttle.
	0	3	17 tret.

Neat 22   2   18 at 19 60 $ cts.
$$22$$

1	$\frac{1}{4}$	39	20
		392	0
	lbs.	431	20
14	$\frac{1}{2}$	9	80
		2	45
2	$\frac{1}{7}$		35
2	$\frac{1}{7}$		35

Ans. $444 15 value.

	C.	qr.	lb.			qr.	lb.	
(3)	4	1	11			1	5	
			6				6	

	C.	qr.	lb.			qr.	lb.	
	26	0	10	gross.	*cwt.*	1	3	2	tare.
	1	3	2	tare.				

$\frac{1}{26}$)24 1 8 suttle.
——0 3 20 tret.

Neat 23 1 16 at 6 75
                     23

qrs.				
1	$\frac{1}{4}$	20	25
		135	0
lbs.		155	25
14	$\frac{1}{2}$	1	$68\frac{3}{4}$
2	$\frac{1}{2}$ $\frac{1}{7}$		$84\frac{1}{4}$
			12

Ans. $157 90 value.

APPLICATION.

	C.	qr.	lb.	
(1)	17	3	22	gross.
		3	14	tare.

Neat 17 0 8=1912 at $23\frac{1}{4}$
                    $23\frac{1}{4}$

5736
3824
478

Ans. $444 54

(2)   *5C. 2qr. 19lb.*   *2qr. 25lb.*

            $3 \times 5 = 15$ 3

     17   0   1   2   0.   19

           5   5

     85   0   5 gross.   C.10   3   11 tare.

     10.   3   11 tare.

Neat 74   0   22 at $6 75*cts.*

                     74

        16 | $\frac{1}{7}$ | 27   00

               472   5

               499   50

        4 | $\frac{1}{4}$ | 96

        2 | $\frac{1}{2}$ | 24

                  12

Ans. $500 82 value.

	C.	qr.	lb.
(3) No. 1.	6	3	18
No. 2.	7	0	10
No. 3.	5	3	26
No. 4.	8	0	3

   lbs.

  8 | $\frac{1}{14}$ | 28   0   1 gross.

  4 | $\frac{1}{2}$ | 2   0   0

          1   0   0

          3   0   0 tare.

Neat 25   0   1 at $3 75*cts.*

                25

            18   75

            75   0

    lb.

    1 | $\frac{1}{112}$ | 93   75

                3   3.

Ans. $93 78 3 value.

(4)   1*C.* 1*qr.* 23*lb.*   18*lb.*
           $4 \times 6 = 24$ 24

    5   3   8   72
         6   36

                   C.qr.lb.
   34   3   20 gross.   432=3 3 12 tare.
    3   3   12 tare.

Neat 31   0   8 at $5 17½*cts.*
                  2

          1035 halves.
           31

          1035
         3105

         32085
            73 9

       2)32168 9
     Ans. $160 79 4 value.

(5)   1*C.* 1*qr.* 13*lb.*   22*lb.*
           $3 \times 5 = 15$ 15

    4   0   11   110
         5   22

                   C.qr.lb.
   20   1   27 gross.   320=2 3 22 tare.
    2   3   22 tare.

Neat 17   2   5 at $9 64*cts.*
                  17

       2 | ⅓ | 67 48
               96 4

       lb. | | 163 38
       4 | 1¼ | 4 82
       1 | ¼ | 34 4
                8 6

    Ans. $169 13 0

```
                     C.  qr.  lb.
        (6)          6   2    14
                               10
     |lbs.|
      16  | ⅐ |66   1    0 gross

       2  | ⅛ | 9   1    24
                  1   0    20
                  ─────────────
                 10   2    16 tare.

      lb.
       4  | 1/26 |55   2    12 suttle.
                   2   0    15 tret.
                  ─────────────
                              lbs.      cts.
        Neat 53   1    25=5989 at 11½
                               11½
                              ─────────
                               65879
                               2994 5
                              ─────────
             Ans. $688 73 5 value.
```

INTEREST.

EXAMPLES IN CASE 1.

```
                 $                         $    cts.
    (2)         225             (3)       384  50
                  7                              5
              ─────────                   ──────────
    Ans. $15 75                  Ans. $19 22 5 m.
              ─────────                   ──────────
```

	£	s.
(4)	580	10
		6

£34 83 0   Ans.

£	s.	d.
34	16	7

20

_____

s.16 60
12
d.7 20

_____

	$	cts.
(5)	1654	81
		5

$82 74 05   Ans.

$	cts.
82	74

(6) |½|½|1500 $

Ans. $7 50

(7) 350 £
      5¼
_____
1750
  87 10
_____
£18 37 10
     20
_____
s.7 50
   12
_____
d.6 00
_____

Ans.

£	s.	d.
18	7	6

(8) |¼|524 $
         5¼
_____
2620
 131
_____
Ans. $27 51

(9) |½|842 $
         5½
_____
4210
 421
_____
Ans. $46 31

CASE 2.

```
              £  s. d.     £  s. d.
$         (3) 124 5 6      4 19 5 Int. for 1 year.
(2)  540               
       5            4           3
     _____    _____    _____
     27|00       £4|97 2 0   £14 18 3 Ans
        2            20
     _____    _____
Ans. $54|00      s.19|42
                    12
                 _____
                 d.5|04
```

```
$
(4)           432
                6
            _____
            $28|92 interest for 1 year
                7
            _____
Ans.    $202|44
```

CASE 3.

```
              $
(2)          325
               4
mo.|  1 |  _____
 2  | 5 | 13|00  Int. for 1 yr.
    |   |  4
    |   | _____
    |   | 52       Int. for 4 yrs.
    |   | 2|16|6   Int. for 2 mo.
    |   | _____
Ans.    $54 16 6
```

(3) $
840
4

mo.		
3	¼	33|60 Int. for 1 *yr.*
		5

168|00 Int. for 5 *yr.*
8|40 Int. for 3 *mo.*

Ans. $176 40

(4) $
840
7

mo.		
4	⅓	58|80 Int. for 1 *yr.*
		5

294|00 Int. for 5 *yrs.*
19|60 Int. for 4 *mo.*

Ans. $313 60

(6) $
1200
5

Ans. $60 00 Int. for 1 *yr.* Then say, as 1*yr.* : 15*w.* ::
$60 : $17 30*cts.* Ans.

(7)

½	½	$240
		4¾
		960
¼	½	120
		60

Ans. $11 40 Int. for 1 *yr.* Then say, as 1*yr.* : 61*d.* ::
$11 40 : $1 90*cts.* Ans.

I

£
(8)   1000
         7
———
£70 00 Int. for 1 yr.   Then as 1yr. : 14mo. :: £70 :
———   £81 13s. 4d   Ans.

$
(9)   450
        5½
———
2250
225
———
$24 75 Int. for 1 yr., Then as 1yr. : 6mo. 20d. ::
——— $24 75cts. : $13 75cts.+ Ans.

$ cts.
(10) 375 25
           6
———
$22 51 50 Int. for 1 yr. Then as 1yr. : 3yrs. 2mo. 3w.
———   5d. :: $22 51cts. 5m. : $72 85. Ans.

CASE 4.

$
(2)   854   (3) $100
        30   48
———   ———
6)25620   8800
———   4400
Ans. $4 27 0   ———
                            6)52800
                            ———
                            Ans. $8 80 0

(4)	$ 3459 75	(5)	$ 1500 60

(4)
$
3459
 75

17295
24213

6)259425

Ans. $43 23 7

(5)
$
1500
 60

6)90000

|1 1/6|15000 m. at 6 per cent.
—2500

Ans. $12 50 0

CASE 5.

(2) 6 yrs.
   4 dolls.

   24 Int. of £100 for 6 yrs.
 +100

 £124 amount of £100 for 6 yrs.

Then as £124 : £1240 :: £100 : 1000. Ans.

(3) 5 yrs.
   6 dolls.

   30 Int. of $100 for 5 yrs.
  100

 $130 amount of $100 for 5 yrs.

Then as $130 : $2470 :: $100 : $1900. Ans.

CASE 6.

(2) $
   1476 amt.
   1200 prin.

   $276 Int.

And $1200 : $100 :: $276 : $23 int. of $100 for the same
 time.
Then as 5 yrs. 9 mo. : $23 :: 1 yr. : $4 per cent. Ans.

$ cts.

(3)   927 82¼ amt.
            834 00   prin.

$93 82½ int.

As $834 : $93 82½cts. :: $100 : $11 25cts.
And then, as 2yrs. 6mo. : $11 25cts. :: 1yr. : $4½ per cent.
   Ans.

CASE 7.

£   £
(2)   1600   2048
             4   1600

£64 00 : 1yr. :: 448 : 7yrs.   Ans.

$
(3)   1000
       4½

40 00
  5 00

$45 00 : 1yr. :: $281 25cts. : 6yrs. 3mo.   Ans.

—••●•••—

COMPOUND INTEREST.

$
(2)   760 prin.
       6 rate per cent.

45 60 int. 1st year.

805 60 amt. of 1st yr. and prin. for the 2d yr.
.48 33 6 int. of 2d yr.

853 93 6 amt. of 2d yr. and prin. for the 3d yr.
  51 23 6 int. of 3d yr.

905 17 2 amt. of 3d yr.
760 00 0 1st prin.

Ans. $145 17 2 compound int.

```
        £  s. d.              £  s. d.
(3)   242 10 6            242 10  6
            6              14 11  0 int. 1st yr.
      _____           _____
      £14|55 3 0          257  1  6 amt.
           20              15  8  5¾ int. 2d yr.
      _____           _____
       11|03              272  9 11¾ amt.
      _____             16  7  0  int. 3d yr.
                         _____
                          288 16 11¾ amt.
                           17  6  7¼ int. 4th yr.
                         _____
                          306  3  7 amt.
                         —242 10  6 1st. prin.
                         _____
                    Ans. 63 13  1+ com. int.
                         _____
```

```
                   $
(4)     1300
              5
        _____
         65|00 int. 1st yr
       1300
        _____
       1365 amt.
          5
        _____
         68|25 int. for 2d yr
       1365
        _____
       1433|25 amt.
          5
        _____
         71|66|2 int. for 3d yr.
       1433|25
        _____
  Ans. $1504 91 2m. amt.
        _____
```

2

(5) $ 3127
              4½

       12508
       1563 5

       $140 71 5

       $ 3127
       140 71 5 int. of the 1st yr

       3267 71 5 amt.
       147   4 7 int. 2d yr.

       3414 .76 2 amt.
       153 66 4 int. 3d yr.

       3568 42 6 amt.
       160 57 9 int. 4th yr.

Ans. $3729 00 5 amt.

PROMISCUOUS EXAMPLES.

(1) $ cts.
       620 25
           5½

       3101 25
       310 12

       34 11 37 int. for 1 yr.
              5

Ans. $170 56 8m.

(2)   £ 420
           7

       £29 40
            20

       s.8 00   Ans. £29 8s

(3) $ 1450
               60

       6)87000

       14500 mills==$14 50cts.   Ans.

	£	s.
(4)	626	5
		5¼

3131 5
156 11 3

£32|87 16 3
20

s.17|56
12

d.6|75
4

qrs.3|00

	£	s.	d.
	626	5	0
	32	17	6¾ int. of the 1st yr.

659 2 6¾ amt.
34 12 1 int. of 2d yr.

693 14 7¾ amt.
36 8 5 int. of 3d yr.

730 3 0¾ amt.
—626 5 0 prin.

Ans. £103 18 0¾+ compound int.

(5)

£
1659
4

Int. for 1 yr. {

£66|36
20

s.7|20
12

d.2|40
4

qr.1|60

Then as 365 *days* : 21 *days* : : £66 7s. 2¼d. : £3 16s. 4¼d.+ Ans.

(6)

$$\frac{\begin{array}{r} \$ \\ 500 \\ 8 \end{array}}{}$$

$$\overline{\$40\ 00}\ \text{int. for 1 yr.}$$

Then as $40 : $500 :: 1y. : 12yrs. 6mo.   Ans.

(7) Thus, 6yrs. and 6mo. at 2 per cent.=$13 interest on $100.

Then $13+$100=$113=amount of $100.

And as $113 : $250 :: $100 : $221 23cts. 9m.   Ans.

(8)
$$\begin{array}{r} £ \\ 450\ \text{amount.} \\ 300\ \text{principal.} \end{array}$$

$$\overline{£150}\ \text{interest.}$$

Then as £300 : £100 :: £150 : £50 which divided by the 5 years=10 per cent. Ans.

→→→●◉●←←←

INSURANCE, COMMISSION AND BROKAGE.

EXAMPLES.

(2)
$$\begin{array}{r} £ \\ 1320 \\ 5 \end{array}$$

Ans. £66|00

(3)
$$\begin{array}{r} \$ \\ 3450 \\ 4\frac{1}{2} \end{array}$$

13800
1725

Ans. $155|25cts.

(4) $\frac{1}{2}|\frac{1}{2}|$
$$\begin{array}{r} \$ \\ 1680 \\ 2\frac{3}{4} \end{array}$$

$\frac{1}{4}|\frac{1}{2}|$
3360
840
420

$46|20 commission.

$1680—$46 20cts.=$1633|80cts.   Ans.

(5) £
     760
       6½
     ―――――
     4560
      380
     ―――――
     £49|40 Ans £49 8s
        20
     ―――――
     s.8|00

(6) ½ |1½| $ 5630
               7¾
          ―――――――
             39410
     ¼ |1½| 2815
             1407 5
     Ans. $436|32|5m.

(7) ½ |1½| $ 17654
                18¾
           ―――――――――
              141232
               17654
     ¼ |1½| 8827
               4413 ½
     Ans. $3310|12 ½

(8) £
     2150
        2
     ―――――
     Ans. £43|00

(9) ¼ |1¼| $ 984 cts. 50
                     1¼
          ―――――――――――
              984 50
              246 12 ½
     Ans. $12|30|62 ½

(10) ½ |1½| $1650 cts. 75
                      1½
           ――――――――――――
               1650 75
                825 37 ½
     Ans. $24|76|12 ½

DISCOUNT.

EXAMPLES.

(2) Thus, *2mo.* at 6 per cent. per an.$=$ $1\frac{1}{2}$ int. of $100
+100
———
$101\frac{1}{2}$ amt. of do.

Then as 101\frac{1}{2}$: $850 :: $100 : $837 43*cts.* 8*m.*+
Ans.

(3) Thus, *9mo.* at 6 per cent. per an.$=$ $4\frac{1}{2}$ int. of $100
100
———
$104\frac{1}{2}$ amt. of 100

Then · as $104\frac{1}{2}$: $645 :: $100 : $617 22*cts.* 4*m.*
present worth   645 00 0
                                    ———
                        Ans. $27 77 6

(4)   *Yrs*
       4
       5
       —
      20 int. of $100 for 4 yrs.
      100
      ———
      $120 amt. of do.

Then as $120 : $775 50*cts.* :: $100 : $646 25*cts.* Ans.

(5) 8*mo.* at 6 per cent. per an.$=$ $4 int. of $100
                                   100
                                   ———
                                   $104 amt. of do.

Then $104 : $580 :: $100 : $557 69*cts.*+. Ans.

Yrs.

(6)

3

$4\frac{1}{2}$

_____

12

$1\frac{1}{2}$

_____

$13\frac{1}{2}$ int. of 100

100

_____

$\$113\frac{1}{2}$ amt. of do.

Then as $\$113\frac{1}{2}$: $\$954$ :: $\$100$: $\$840$ 52*cts.* 8*m.* Ans.

(7) Thus, 15 *mo.* $= 1\frac{1}{4}yr.$ at 7 per cent. per an-
num$=\$8\frac{3}{4}$ the discount of 100.

100

_____

$\$108\frac{3}{4}$ amt.

Then $\$108\frac{3}{4}$: $\$205$ :: $\$100$: $\$188$ 50*cts.* 5*m.* present worth.   205 00

_____

Ans. $\$16$ 49 5

mo.   £

(8) | 6 | $\frac{1}{2}$ | 5

| 3 | $\frac{1}{2}$ | $2\frac{1}{2}$

$1\frac{1}{4}$

_____

$3\frac{3}{4}$ discount of 100

100

_____

$\$103\frac{3}{4}$ amt.

Then as $\$103\frac{3}{4}$: $\$775$ :: $\$100$: $\$746$ 98*cts.* 7*m.* Ans.

(9)

mo.	£		
6	1½	6	
3	1⅔	3	
1	1⅓	1½	1½

Again

mo.	£	
3	¼	6
		1 qr.

6
1 r½

15 mo.

5 dis. of 100 for 10 mo.   7½ dis. of 100 for

100   100

$105 amt.   107½

$
1005
—475

Rem. 530

Then as 105 : 475 :: 100 : 452 38.   **Ans. to first part.**
Again 107½ : 530 :: 100 : 493 02 4

       Ans. $945 40 4m.

(10) $
     2260
       6

   Again $
         6
         5

135 60 int. for 1 yr.   30 dis. of 100
    5   100

$678 00 int. for 5 yrs. $130 amt.

Then $130 : $2260 :: $100 : $1738 46cts. 2m. pres. wr.
                  2260 00 0

                 521 53 8 discount.
                 678 00 0 interest.

       Ans. $156 46 2

(12) £782
        4
      ─────
     £31|28
        20
      ─────
     s.5|60
       12
     d.7|20
     ─────

(13) $476
        3
      ─────
   Ans. $14|28cts.
      ─────

Ans. £31 5s. 7d.

(14) $1385
         6
      ─────
       83 10 dis.
     1385 00
     ─────
   Ans. $1301 90cts.

(15) $650
        4½
      ─────
      2600
       325
      ─────
     29|25 discount.
    650|00
     ─────
   Ans. $620|75

EQUATION.

EXAMPLES.

(2) $250 × 6 = 1500
     250 × 8 = 2000
     ─────
      500   3500 ÷ 500 = 7mo. Ans.

£

(3)
$$100 \times 2 = 200$$
$$100 \times 4 = 400$$
$$100 \times 6 = 600$$

300       $1200 \div 300 = 4mo.$ Ans.

$

(4)
$$100 \times 3 = 300$$
$$200 \times 5 = 1000$$
$$250 \times 8 = 2000$$

550       $3300 \div 550 = 6mo.$ Ans.

BARTER.

EXAMPLES.

(1) Thus 2cwt. 2qrs. 13lbs.=293lbs. × 9cts.=2637cts.
Then as 25cts. : 2637cts. : : 1lb. . 105lbs. $7\frac{17}{25}oz.$ Ans.

(2) Thus 2500lbs. × $4\frac{1}{2}$cts.=$112 50cts.
Then as $1 30cts. : $112 50cts. : : 1lb. : 86lbs. 8oz.+
Ans.

(3) Thus 108lbs. × $1 25cts.=$135 00cts.
Then as $8\frac{3}{4}$cts. : $135 00cts. : : 1lb. : 1542lb. 13oz.+
Ans.

(4) First, as 1cwt. : $3 75cts. : : 14cwt. 3qrs. 26lbs. : $56
18cts. 3m. the value of the rice.
Then as $1 $87\frac{1}{2}$cts. : $56 18cts. 3m. : : 1lb. : 29lbs.
15oz.+ Ans.

(5) Thus 2cwt. 3qrs. 17lbs.=325lbs. × $12\frac{1}{2}$cts.=$40 $62\frac{1}{2}$cts.
Then as 37cts. : $40 $62\frac{1}{2}$cts. : : 1yd. : 109yds. 3qrs.
Ans.

(6) Thus 357bu. × 93cts.=$332 01ct.
Then 45cts. : $332 01ct. : : 1bu. : 737bu. 3pe.+ Ans.

(7) Thus 15*cwt.* 0*qr.* 27*lbs.* = 1707*lbs.* × 20*cts.* = $341 40*cts.*
Then $9 50*cts.* : $341 40*cts.* :: 1*cwt.* : 35*cwt.* 3*qrs.* 20*lbs.*+ Ans.

(8) Thus 95*yds.* × 5*pie.* = 475*yds.* × 23*cts.* = $109 25*cts.*
And 32 *sheep* × 250 = ———————— —80 00

                                    $29 25 rem.

Then as $1 50*cts.* : $29 25*cts.* :: 1*cwt.* : 19*cwt.* 2*qrs.* Ans.

(9) Thus 1286*yds.* at 43*cts.* per *yd.* = $552 98*cts.*
And 2*cwt.* 1*qr.* 13*lbs.* = 265*lbs.* × 14*cts.* = 37 10—

                              Ans. $515 88

(10) Thus 570*lbs.* × 7*cts.* = $39 90*cts.*
Then as 11½*cts.* : $39 90*cts.* :: 1*lb.* : 346*lbs.* 15*oz.*+ Ans.

(11) Thus 112*cwt.* × $5 04*cts.* = $564 60*cts.*
Then as 1208*yds.* : $564 60*cts.* :: 1*yd.* : 46*cts.* 7*m.*+ Ans.

(12) Thus 750*lbs.* × $1 08*cts.* = $810 00*cts.*
Then 8*cts.* : $810 00*cts.* :: 1*lb.* : 10125*lbs.* = 90*cwt.* 1*qr.* 17*lbs.* Ans.

(13) Thus 2*hhds.* = 126*gals.* × 75*cts.* = $94 50*cts.*
Then 56*yds.* : $94 50*cts.* :: 1*yd.* : $1 68¾*cts.* Ans.

(14) Thus 2108*lbs.* × 10*cts.* = $210 80*cts.*
And 31*doz.* × 11½*cts.* = +3 56½

                      $214 36½ amt. of the whole.
                      —135 25

                      $79 11½ rem.

Then as $1 58*cts.* : $79 11½*cts.* :: 1*bu.* : 50*bar.*+ Ans.

5) Thus $17cwt. \times 4 \times 28 = 1904lbs. \times 13\frac{1}{2}cts. = \$257\ 04cts.$ value of A.'s goods.

And $1200lbs.$ at the rate of $14 per cwt.$ $= 150\ 00$ balance of B.'s goods.

Ans. A. is to receive $107 04

6) Thus 25cts.

—20

—

5 gain on 20cts.

—

Then as $5cts. : 20cts. :: 5cts. : 20cts.$ Ans.

7) Thus $50cts. : 56cts. :: 31\frac{1}{4}cts. : 35cts.$ Ans.

8) Thus 105 *tons* at $10 03 per ton $= \$1053\ 15cts.$ value of the iron.

pays cash	650 00
250*lbs.* at 20cts. per *lb.*=	50 00
10 loads $\times$ 15*bu.* $\times$ 45cts.=	67 50
And 85*gals.* at the rate of $75 per *hhd.*=	101 19
	—868 69
	1053 15

Rem. unpaid $184 46

Then $30cts. : \$184\ 46cts. :: 1lb. : 615lbs.$ nearly. Ans.

—◉—

LOSS AND GAIN.

(2)   Thus 10cts.

—8

—

2

—

Then $1lb. : 1763lbs. :: 2cts. : \$35\ 26cts.$ Ans.

(3)   Thus $5 25cts.
            —5 00
            ———
            25 gained per barrel.
            ———

Then 1bar. : 363bar. :: 25cts. : $90 75cts.   Ans.

(4)   Thus $3 90cts.
            —3 75
            ———
            15 gained per yard.
            ———

Then 1yd. : 150yds. :: 15cts. : $22 50cts.   Ans.

(5) First, 1cwt. : $7 50cts. :: 18cwt. 2qrs. : $138 75cts.
the cost.
Then 1cwt. : $7 75cts. :: 18cwt. 2qrs. : $143 37½cts.
sold for.
                               Ans. gained $4 62½

(6) First, 210 reams × $2 62½ = $551 25cts. the cost.
And   210 reams × $2 87½ = $603 75cts. sold for.
                           Ans. $52 50 gained.

(7)   Thus, sold for $20 75cts.
             cost   18 12½
             ————
             gained $2 62½ Ans.

(8)   First, 50cts.
            —45
            ——
            5
            ——

Then 1bu. : 150bu. :: 5cts. : $7 50cts.   1st Ans.
Again, 50cts. : 5cts. :: $100 : $10.   2d Ans.

9) First, 760*lbs.*×90*cts.*=$684 00 sold for.
810 00 cost.
_____
Lost 126 00 1st Ans.

Then $810 : $126 :: $100 : $15⅝. Ans.

(10) First, 37½*cts.*
32
_____
5½

Then 37½*cts.* : 5½*cts.* :: $100 : $14⅔ per cent. Ans.

11) Thus 1*s.* : 2*d.* :: £100 : £16⅔ per cent. Ans.

12) Thus $13 75*cts.* First cost of each piece.
3 12½ for dyeing.
_____
$16 87½ whole cost.

Then $100 : $112 :: $16 87½*cts.* : $18 90*cts.* Ans.

13) Thus 1*cwt.* : 1*lb.* :: $7+$3 : 8*cts.* 9*m.* Ans.

14) Thus, paid 23*cts.* per *lb.*
Sold it for 19
___
Lost 4*cts.* per *lb.*
___
Then as 1*lb.* : 702*lbs.* :: 4*cts.* : $28 08*cts.* Ans.

(15) Thus $2 23*cts.* : $2 75*cts.* :: $110 : $135 65*cts.*
And $135 65*cts.*—$100=$35 65*cts.*=35⅔ nearly.
Ans.

(16) Thus $100 : $125 :: $2 10*cts.* : $2 62½*cts.* what
1 box sold for.
Then as $3 50*cts.* price of 1*cwt.* : $2 62½*cts.* price
of 1 box :: 112*lbs.* : 84*lbs.* Ans.

(17)   First, 16*pie.* × $14=$224 the prime cost.
And 5*pie.* × $17=$85
      6*pie.* × $15=$90
                  —————
            $175 received back again.

Then as $100 : $112 :: $224 : $250 88*cts.* price of
the whole with rate per cent. added.—175 00

                        5)75 88 price of the
                        —————— 5 pieces.
                        Ans. $15 17 6 per *pie.*

(18) Thus $500—$410=$90 gain on the whole.
Then as 372*lbs.* : 1*lb.* :: $90 : 24*cts.* 1*m.*+ Ans.

(19) Thus $1 : $100 :: 5*cts.* : $5 00 the Ans.

(20) First, $1 05*cts.* ×510=$535 50*cts.* prime cost.
And $1 30*cts.* ×510=$663 00*cts.* sold for.

             mo. $
            |3|¼| 6
            | | ————
            | | 1 50
            | |100 00
            ————————————
             $101 50
            ————————————

Then $101 50*cts.* : $100 :: $663 : $653 20*cts.*+
Hence $653 20*cts.*—$535 50*cts.*=$117 70*cts.* Ans.

FELLOWSHIP.

EXAMPLES.

CASE 1.

(2)　　　Thus D 's stock $500
　　　　　　　E.'s　　　　400
　　　　　　　F.'s　　　　300
　　　　　　　　　　　　　——
　　　　　　　Sum　1200

　　　　　$　　　$　　　$　　　$
Then as 1200 : 500 :: 300 : 125=D.'s ⎫
And　　1200 : 400 :: 300 : 100=E.'s ⎬ Ans.
And　　1200 : 300 :: 300 :　75=F.'s ⎭

(3)　　　　Thus A. $1200
　　　　　　　　B.　 500
　　　　　　　　C.　 700
　　　　　　　　　　 ——
　　　Whole debt $2400

　　　　　$　　　$　　　$　　　$
Then as 2400 : 1200 :: 1800 : 900 A.'s ⎫
　　as 2400 :　500 :: 1800 : 375 B.'s ⎬ Ans
　　as 2400 :　700 :: 1800 : 525 C.'s ⎭
　　　　　　　　　　　　　　　———
　　　　　　　　　　　　　$1800 proof.

(4)　　　Thus A. had 50 *cattle*.
　　　　　　　 B.　 · 80
　　　　　　　 C.　　70
　　　　　　　　　　 ——
　　　　　　　Sum 200

　　cattle. cattle.　$　　$
Then as 200 : 50 :: 60 : 15 A.'s ⎫
　　as 200 : 80 :: 60 : 24 B.'s ⎬ Ans.
　　as 200 : 70 :: 60 : 21 C.'s ⎭
　　　　　　　　　　　　——
　　　　　　　　　 $60 proof.

(5)   Thus, to A. $120
               B. 250 75
               C. 300
               D. 208 25
               ───────
           Sum 879 00

Then
As $879 : $650 { ∷ 120   : 88 75+ =A.'s sh.
                 ∷ 250 75 : 185 42+ =B.'s sh.   } Ans.
                 ∷ 300   : 221 84+ =C.'s sh.
                 ∷ 208 25 : 153 99+ =D.'s sh. }

(6)   Thus A. is to have 1 portion.
               B.   2
               C.   6
                            ──
                            9 sum of the portions.

Then as { 9 : 1 ∷ 900 : 100=A.'s share.
          9 : 2 ∷ 900 : 200=B.'s share.   } Ans.
          9 : 6 ∷ 900 : 600=C.'s share. }

(7)   Thus, he owes to A. 250 50
                         B. 500 00
                         C. 349 50
                         ────────
                       Sum 1100 00

Then $ cts. $ cts. m.
As 1100 : 960 ∷ { 250 50 : 218 61 8+A.'s
                  500 00 : 436 36 3+B.'s } Ans.
                  349 50 : 305 01 8+C.'s }

EXAMPLES

CASE 2.

(1)   Thus $\$$
$88 \times 3 = 264$
$120 \times 4 = 480$
$300 \times 6 = 1800$

Sum of stocks and time 2544

$$\text{Then as } \$2544 : \begin{cases} \$ \quad\quad \$ \quad\quad \$ \; cts.\, m. \\ 264 :: 184 : \; 19 \; 09 \; 4 = \text{L.'s} \\ 480 :: 184 : \; 34 \; 71 \; 6 = \text{M.'s} \\ 1800 :: 184 : 130 \; 18 \; 8 = \text{N.'s} \end{cases} \text{Ans.}$$

(2) $\$ \quad m. \quad \$$
$580 \times 3 = 1740$
$+100$
——
$680 \times 9 = 6120$
——
A.'s product 7860
——

$\$ \quad m. \quad \$$
$1000 \times 9 = 9000$
$+200$
——
$1200 \times 3 = 3600$
——
B.'s product 12600
——

$\$ \quad m. \quad \$$
$480 \times 3 = 1458$
-300
——
$180 \times 2 = 372$
$+500$
——
$686 \times 3 = 2058$
-400
——
$286 \times 1 = 286$
$+1000$
——
$1286 \times 3 = 3858$
——
C.'s product 8032
——

A.'s   7860
B.'s   12600
C.'s   8032
——

$$28492 : 2108 \; 44 :: \begin{cases} \$ \quad\quad \$ \; cts.\, m. \\ 7860 : 581 \; 64 \; 8 + A \\ 12600 : 932 \; 41 \; 4 + B \\ 8032 : 594 \; 37 \; 7 + C \end{cases} \text{Ans.}$$

EXCHANGE.

DOMESTIC EXCHANGE.

(1) Thus, £63 14*s*. 6*d*.=15294*d*.÷72*d*. a dollar in Virginia=\$212 41½*cts*. Ans.

(2) Thus, £230 10*s*. 7*d*.=55327*d*.÷96*d*. a dollar in New-York and N. Carolina=\$576 32*cts*. 2*m*. Ans.

(3)   Thus, \$150

     90*d*.=a doll. Penn. cur.
   12)13500*d*.
   —————

   2|0)112|5
   —————

   £56 5*s*. Ans.

(4)   Thus, \$377 40*cts*.
       72*d*.=a doll. Mass. cur.
    754 80
   26418 0
   —————
   12)27172 80
   —————
   2|0)226|4 4*d*.
   —————
   £113 4*s*. 4*d*. Ans.

(5)   Thus, \$389 45*cts*.
      56*d*.=a doll. in Georgia.
   —————
    233670
    194725
   —————
   12)21809|20
   —————
   2|0)181|7 5
   —————
   £90 17*s*. 5*d*.

FOREIGN EXCHANGE.
EXAMPLES.

(2) Thus £1 : £76 :: $4 10*cts.*=£1 Irish : $311 60 *cts.* Ans.

(3) Thus $1 24*cts.* = 1 milrea : $532 33*cts.* :: 1*m.* : 429*m.* 298*reas.*+ Ans.

(4) Thus 66*cts.* : $1869 :: 1*ru.* : $2831\frac{9}{11}$*ru.* Ans.

(5) Thus 1*g.* :, 165*g.* :: 39*cts.* : $64 35*cts.* Ans.

(6) Thus 33*cts.* 5*m.*=1*m. b.* : $280 58*cts.* 5*m.* :: 1*m. b.* : 837*m. b.*+ Ans.

(7) Thus 1*li.* : 562*li.* :: 18*cts.* 5*m.*=1*li.* : $103 97*cts.* Ans.

(8) Thus 10*cts.*=1*rial plate* : $463 :: 1*rial* : 4630*rials.* Ans.

(9) Thus 1*flo.* : 40*cts.* :: 591*flo.* 17*st.* : $236 74*cts.*
Or 1*st.* : 2*cts.* :: 591*flo.* 17*st.* : $236 74*cts.*
Then $100 : $160 :: $236 74*cts.* : $378 78*cts.*+ Ans.

(10) Thus as 100*cr.*+25 ; 100*b.* :: 2464*m. b.* : 1971*m. b.* 3*sch.* $2\frac{3}{4}$*pen.* Ans.

(11) Thus 1*cr.* : $32\frac{1}{2}$*d.* :: 2000*cr.* : £270 16*s.* 8*d.* Ans.

(12) Thus as 1*pi.*=8*ri.* : 36*d.* :: $1676 6*ri.*=16766*ri.* : £314 7*s.* 3*d.* Ans.

(13) Thus 1*pez.*=20*sol.* : 54*d.* :: 3940*pez.* 15*sol.* : £886 13*s.* $4\frac{1}{2}$*d.* Ans.

(14) Thus 1*ru.* : 4*s.* 3*d.* :: 2586*ru.* : £549 10*s.* 6*d.* Ans.

(15) First £1 : £450 15*s.* :: 34*s.* 6*d* 186610$\frac{1}{2}$*pence.*
Or 20*s.* : 9015*s.* :: 414*d.* : 186610$\frac{1}{2}$*pence* Flemish, or groots.
Then 50*st.*=100*d.* : 186610$\frac{1}{2}$*d.* :: 1*ru.* : 1866*ru.* 10$\frac{1}{2}$ *cop.* Ans.

(16) Thus as £108 6*s.* 8*d.* Irish : £100*str.* :: £813 3*s.* 6*d.* : £750 12*s.* 6*d.* Sterling. Ans.

(17) First 20s. : 33s. 6d. :: 5s. : 8s. $4\frac{1}{2}d$.
Then 5s. : 8s. $4\frac{1}{2}d$. : : $32\frac{1}{2}d$. : $54\frac{7}{16}d$. Flemish. Ans

(18) Thus $32\frac{1}{2}d$. : $54\frac{7}{16}d$. : : 5s. : 8s. $4\frac{1}{2}d$.
Then as 5s. : 8s. $4\frac{1}{2}d$. : : $20_0{}^{s}$: $3\frac{3s}{3}$. 6d. **Ans.**

(19) Thus $\begin{array}{c} s. \\ |5| \end{array} \begin{array}{c} s.\ d. \\ \frac{1}{4}|33\ 6 \end{array}$

       8 $4\frac{1}{2}$=value of a crown at that rate.
Then 8s. $4\frac{1}{2}d$. : 5s. : : $54\frac{7}{16}d$. : $32\frac{1}{2}d$. **Ans.**

(20) Thus $32\frac{1}{4}d$. : 32d. : : 36s. 6d. : 36s. $2\frac{20}{13}d$. **Ans.**

(21) Thus 51d. : 53d. : : 42d. : $43\frac{11}{17}d$. **Ans.**

—••••—

VULGAR FRACTIONS.

REDUCTION OF VULGAR FRACTIONS.

EXAMPLES.
CASE 1.

(2)   Numer. 108)144(1
                108

Common measure 36)108(3
                  108

Then $36)\frac{108}{144}=\frac{3}{4}$. **Ans.**

(4)   Numer. 126)234(1
                126

           108)126(1
              108

Common measure 18)108(6
                 108

Then $18)\frac{126}{234}=\frac{7}{13}$. **Ans.**

L

CASE 2.

(2) $\qquad 45 \times 3 + 2 = 1\frac{37}{3}$. Ans.

(3) $\qquad$ Thus $1564 \times 5 + 3 = 7\frac{823}{5}$. Ans.

CASE 3.

(2) $\qquad$ Thus $67 \div 7 = 9\frac{4}{7}$. Ans.

(3) $\qquad$ Thus 16)364($22\frac{12}{16}$. Ans.

$$\begin{array}{r} 32 \\ \hline 44 \\ 32 \\ \hline 12 \\ \hline \end{array}$$

CASE 4.

(2) $\qquad$ Thus $6 \times 8 \times 11 \times 13 = 6864$ numer.
And $7 \times 9 \times 12 \times 17 = \overline{12852}$ denom. $= \frac{572}{1071}$. Ans.

(3) $\qquad$ Thus $7 \times 15 \times 8 \times 6 = 5040$ numer.
And $12 \times 19 \times 11 \times 13 = \overline{32604}$ denom. $= \frac{420}{2717}$. Ans.

CASE 5.

(2) $\qquad$ Thus 5)5 20 10 15 the denominators.

$$\begin{array}{r} 2)1 \quad 4 \quad 2 \quad 3 \\ \hline 1 \quad 2 \quad 1 \quad 3 \\ \hline \end{array}$$

Then $5 \times 2 \times 1 \times 2 \times 1 \times 3 = 60$ common denom.
Then the com. denom.
$$\left. \begin{array}{l} 60 \div 5 = 12 \times 4 = 48 \\ 60 \div 20 = 3 \times 9 = 27 \\ 60 \div 10 = 6 \times 7 = 42 \\ 60 \div 15 = 4 \times 4 = 16 \end{array} \right\} \text{numer.}$$

That is $\frac{48}{60}$ $\frac{27}{60}$ $\frac{42}{60}$ $\frac{16}{60}$. Ans.

(3)   Thus 2)10 2 9 the denom.

$$5\ 1\ 9$$

Then $2\times5\times1\times9=90$ common denom.

$$90\div10=\ 9\times9=81$$
$$90\div\ \ 2=45\times1=45$$
$$90\div\ \ 9=10\times5=50$$
$\Big\}$ numer.

That is $\frac{81}{90}$ $\frac{45}{90}$ $\frac{50}{90}$. Ans.

CASE 6.

(2) First 1*lb.* troy$=240dwt.$ therefore $\frac{3}{8}$ of $\frac{1}{240}=\frac{3}{1920}=\frac{1}{400}lb.$ Ans.

(3) Thus $\frac{3\times1\times1}{8\times4\times4}=\frac{3}{128}$. Ans.
And

(4) Thus 1*hhd.*$=504pts.$ therefore $\frac{5}{8}$ of $\frac{1}{504}=\frac{5}{4032}hhd.$ Ans.

(5) Thus 8*fur.*$=1m.$ therefore $9\times1=9$ the numer. and $16\times8=128$ the denom.$=\frac{9}{128}$. Ans.

CASE 7.

(2) Thus $2\times112=224$ the numer. and $252\times1=252$ the denom.$=\frac{224}{252}=\frac{8}{9}lb.$ Ans.

(3) $\frac{6}{1080}$ of £1$=\frac{6}{1080}$ of $\frac{240}{1}=\frac{1440}{1080}=\frac{6}{7}d.$ Ans.

CASE 8

(2) Thus $\frac{7}{8}$ of a shilling$=\frac{7}{8}$ of $\frac{12}{1}=\frac{84}{8}=10\frac{1}{2}d.$ Ans.

(3) Thus $\frac{12}{48}$ of a day$=\frac{12}{48}$ of $\frac{24}{1}=\frac{288}{48}=6hrs.$ Ans.

(4) Thus $\frac{5}{16}$ of an acre$=\frac{5}{16}$ of $\frac{4}{1}$ of $\frac{40}{1}=\frac{800}{16}$ perches$=$ 1*r.* 10*p.* Ans.

CASE 9.

(2) Thus 5*s.* 4*d.*$=64d.$ and £1$=240d.$ therefore $\frac{64}{240}=\frac{4}{15}$£. Ans.

(3) Thus $6mo.$ $2w.=26w.$ and $1yr.=52w.$ therefore $\frac{26}{52}$ of $1yr.=\frac{1}{2}yr.$ Ans.

(4) Thus $2qrs.$ $3na.=11na.$ and $1yd.=16na.$ therefore $\frac{11}{16}yd.$ is the Ans.

ADDITION OF VULGAR FRACTIONS.

EXAMPLES.

(2) Thus $\frac{3}{13}+\frac{4}{13}+\frac{5}{13}+\frac{1}{13}=\frac{13}{13}=1$. **Ans.**

(3) Thus $\frac{4}{7}+\frac{3}{7}+\frac{6}{7}=\frac{13}{7}=1\frac{6}{7}$. Ans.

(4) Thus 5)5 10

       1　2=10 common denom.

And $10\div 5\times 2=4$ ⎱ numer.

     $10\div 10\times 5=5$ ⎰

Whence $\frac{4}{10}+\frac{5}{10}=\frac{9}{10}$. Ans.

(5) Thus $3\frac{1}{4}=\frac{13}{4}$, $8\frac{2}{7}=\frac{58}{7}$, and $4\times7\times9=252$ common denom.

And $252\div4\times13=\ \ 819$ ⎱

   $252\div7\times58=2088$ ⎰ numer.

   $252\div9\times\ \ 4=\ \ 112$ ⎰

Whence $\frac{819}{252}+\frac{2088}{252}+\frac{112}{252}=\frac{3019}{252}=11\frac{247}{252}$. Ans.

(6) Thus $\frac{3}{8}$ of $\frac{5}{6}=\frac{15}{48}=\frac{5}{16}$, and $\frac{2}{4}$ of $\frac{7}{12}=\frac{14}{48}=\frac{7}{24}$. Then 8)16 24

      2　3=48 common denom.

And $48\div16\times5=15$ ⎱ numer.

    $48\div24\times7=14$ ⎰

Whence $\frac{15}{48}+\frac{14}{48}=\frac{29}{48}$. Ans.

(7) Thus $\frac{1}{3}$ of $\frac{4}{7}$ of $\frac{40}{1}=\frac{160}{3}=53\frac{1}{3}per.=1r.$ $13\frac{1}{3}p.$

And $\frac{7}{10}$ of $\frac{40}{1}=\frac{280}{1}=28p.$

Whence $1R.$ $13\frac{1}{3}p.$

     0　28

Ans. 2　1$\frac{1}{3}$

MULTIPLICATION OF VULGAR FRACTIONS.

EXAMPLES.

(2) $\frac{2}{10}$ by $\frac{1}{3}$ thus $\dfrac{2\times1=2}{10\times3=30}=\frac{1}{15}$. Ans.

(3) Thus $6\frac{2}{4}=\frac{26}{4}$ by $\frac{1}{7}=\dfrac{26\times1=26}{4\times7=28}=\frac{13}{14}$. Ans.

(4) $4\frac{3}{4}=\frac{19}{4}$ by $\frac{2}{3}=\dfrac{19\times2=38}{4\times3=12}=3\frac{2}{12}=3\frac{1}{6}$. Ans.

SUBTRACTION OF VULGAR FRACTIONS.

EXAMPLES.

(2) Thus $\frac{1}{7}$ of $\frac{1}{4}=\frac{1}{28}$ whence $\frac{19}{20}-\frac{1}{28}$.

  4)20 28

     5   7$=140$ common denom.

  $140\div20\times19=133$ ⎰

  $140\div28\times\ 1=\ \ \ 5$ ⎱ numer.

  Whence $\frac{133}{140}-\frac{5}{140}=\frac{128}{140}=\frac{32}{35}$. Ans.

(3) Thus $1\times14=14$ common denom.

  And $14\div\ 1\times5=70$ ⎰

      $14\div14\times8=\ \ 8$ ⎱ numer.

  Whence $\frac{70}{14}-\frac{8}{14}=\frac{62}{14}=4\frac{6}{14}$. Ans.

(4) Thus $\frac{2}{3}$ of a league$=\frac{2}{3}$ of 3 miles$=2$ miles.

  And $\frac{7}{10}$ of a mile$=\frac{7}{0}$ of 8 furlongs$=\frac{56}{10}=5\frac{6}{10}$ furlongs$=5$ furlongs 24 poles.

  Therefore $2m.-5fur.\ 24po.=1m.\ 2fur.\ 16po.$ Ans.

(5) Thus $5\frac{3}{4}=\frac{23}{4}$ and $2\frac{2}{3}=\frac{8}{3}$ therefore $4\times3=12$ com. d.

  And $12\div4\times23=69$ ⎰

     $12\div3\times\ 8=32$ ⎱ numer.

  Whence $\frac{69}{12}-\frac{32}{12}=\frac{37}{12}=3\frac{1}{12}$. Ans.

(6) Thus $\frac{2}{3}$ of $\frac{7}{10}=\frac{14}{18}$ and $\frac{1}{4}$ of $\frac{3}{5}=\frac{3}{20}$.

  And 4)48 20

     12 5$=240$ common denom.

  And $240\div48\times14=70$ ⎰

     $240\div20\times\ 3=36$ ⎱ numer.

  Whence $\frac{70}{240}-\frac{36}{240}=\frac{34}{240}=\frac{17}{120}$. Ans.

DIVISION OF VULGAR FRACTIONS.

EXAMPLES.

(2) $\frac{5}{8}$ by $\frac{3}{1}$ thus $\frac{1}{3})\frac{5}{8}(\frac{5}{24}$. Ans.

(3) $6\frac{3}{5}=\frac{33}{5}\div\frac{1}{3}$ thus $\frac{3}{1})\frac{33}{5}(\frac{99}{5}=19\frac{4}{5}$. Ans.

(4) Thus $\frac{1}{2}$ of $\frac{3}{4}=\frac{6}{12}$ and $\frac{1}{2}$ of $\frac{2}{3}=\frac{2}{6}$.
Then $\frac{6}{12}\div\frac{2}{6}$ thus $\frac{6}{2})\frac{6}{12}(\frac{36}{24}=1\frac{1}{2}$ Ans.

(5) $\frac{1}{8}$ by $\frac{3}{4}$ thus $\frac{4}{3})\frac{1}{8}(\frac{4}{18}=\frac{2}{9}$. Ans.

(6) $\frac{2}{8}$ of $\frac{7}{8}=\frac{14}{24}$ and $\frac{1}{4}$ of $\frac{1}{4}=\frac{1}{26}$.
Then $\frac{14}{34}$ by $\frac{1}{28}$ thus $\frac{28}{1})\frac{14}{24}(\frac{392}{24}=16\frac{1}{3}$. Ans.

(7) $\frac{1}{2}$ of $17\frac{1}{2}=\frac{1}{2}$ of $\frac{35}{2}=\frac{35}{4}$.
Then $\frac{35}{4}\div\frac{3}{4}$ thus $\frac{4}{3})\frac{35}{4}(\frac{140}{12}=11\frac{2}{3}$. Ans.

(8) Thus $\frac{3}{4}$ of $91\frac{78}{97}=\frac{3}{4}$ of $\frac{8905}{97}=\frac{26715}{388}$.
And $\frac{26715}{388}\div\frac{1036}{50}$ thus $\frac{50}{1036})\frac{26715}{388}(\frac{1335750}{401968}=3\frac{64923}{200081}$. Ans.

RULE OF THREE IN VULGAR FRACTIONS.

EXAMPLES.

(2) Thus $3\frac{1}{4}yds.=\frac{13}{4}$ and $9\frac{3}{4}s.=\frac{39}{4}$ and $4\frac{3}{4}yds.=\frac{19}{4}$.
Then we have $\frac{13}{4}:\frac{39}{4}::\frac{19}{4}:14s.\ 3d.$
For $\frac{39}{4}\times\frac{19}{4}=\frac{741}{16}s.\div\frac{4}{13}=\frac{3964}{208}=14s.\ 3d.$ Ans.

(3) Thus $\frac{5}{4}:\frac{20}{1}::\frac{3}{4}:12yds.$
For $\frac{20}{1}\times\frac{3}{4}=\frac{60}{4}\div\frac{4}{5}=\frac{240}{20}yd.=12yds.$ Ans.

(4) Thus $27\frac{3}{4}\times4pe.=111yds.$ and $15\frac{2}{3}s.=15s.\ 8d.$
Then say as in whole numbers, $1yd.:111yds.::15s.$
$8d.:£86\ 19s.$
For $15s.\ 8d.=188d.\times111yds.=20868d.$ which$\div12$
$\div20=£86\ 19s.$ Ans.

(5) Thus $5\frac{3}{7}cwt.=\frac{38}{7}$ and $£31\frac{14}{32}=\frac{1006}{32}$.
Then we have $\frac{38}{7}:\frac{2}{5}::\frac{1006}{32}:£2\ 6s.\ 3\frac{14}{16}d.$
For $\frac{1006}{32}\times\frac{2}{5}=\frac{2012}{160}\div\frac{7}{38}=\frac{14084}{6080}£.=£2\ 6s.\ 3\frac{14}{16}d.$ Ans.

(6) First $1\frac{2}{3}lb.=\frac{8}{3}$.

Then $\frac{1}{3}lb. : \frac{5}{3} :: \frac{4}{7}dol. : \$2\ 74\frac{2}{7}cts.$

For $\frac{8}{3}\times\frac{4}{7}=\frac{32}{33}\div\frac{3}{1}=\frac{96}{33}dol.=\$2\ 74\frac{2}{7}cts.$ Ans.

(7) Thus $20\frac{2}{3}d.=\frac{62}{3}$.

Then inversely thus $6m. : 10m. :: \frac{62}{3}day. : 34\frac{4}{9}days.$

For $\frac{62}{3}\times\frac{10}{1}=\frac{620}{3}\div\frac{1}{6}=\frac{620}{18}=34\frac{4}{9}days.$ Ans.

(9) First $\frac{1}{3}$ of $2\frac{1}{2}cwt.=\frac{1}{3}$ of $\frac{5}{2}=\frac{5}{6}$ of a $cwt.$

Then this reduced to $lbs.$ would be $\frac{5}{6}$ of $\frac{112}{1}=\frac{560}{6}$.

Then we have $6\frac{1}{2}lbs.=\frac{13}{2} : \frac{560}{6} :: \frac{3}{4} : \$10\ 76\frac{12}{13}cts$

For $\frac{560}{6}\times\frac{3}{4}=\frac{1680}{24}\div\frac{2}{13}=\frac{3360}{312}dol.=\$10\ 76\frac{12}{13}cts.$

—➤●◉●◄—

DECIMAL FRACTIONS.

ADDITION OF DECIMALS.

EXAMPLES.

(5)
```
      56.12
        .7
      1.314
    5837.01
        .15
    ----------
Ans. 5895.294
```

(6)
```
     361.04
       .120
     78.0006
     101.54
       8.943
        .3
    ----------
Ans. 549.9436
```

MULTIPLICATION OF DECIMALS.

EXAMPLES.

(2)
```
     54.20
     38.63
    --------
     16260
     32520
     43360
     16260
    ----------
Ans. 2093.7460
```

(3)
```
     4560.
      .3720
    --------
     91200
     31920
     13680
    ----------
Ans. 1696.3200
```

(4)   .28043
           .0005
           _____

Ans. .000140215

SUBTRACTION OF DECIMALS.
EXAMPLES.

(5)   13.16421   (6)   5960.
        4.286   .3742
       _____   _____
Ans.   8.37821   Ans. 5959.6258

DIVISION OF DECIMALS.
EXAMPLES.

(2) 4 20)148.63(35.304+   Ans.
       1263

       2233
       2105
       ____
       1200
       1263

        1700
        1684
        ____
        16 rem.

(4)   931)2.00385(.0021523+ Ans.
       1862
       ____
       1418
        931
       ____
       4875
       4655
       ____
       2200
       1862
       ____
       3380
       2793
       ____
       587 rem.

(3) 3.2).2142(.066+   Ans.
       192

       222
       192
       ___
       30 rem.

REDUCTION OF DECIMALS.

CASE 1.

(2) 8)7.000
```
   .875  Ans.
```

(3) 24)170(.70833+
```
     168
     ───
     200
     192
     ───
      80
      72
     ───
      80
      72
     ───
       8 rem.
```

(4) 2162.)3810(.1762+ Ans.
```
    2162
    ─────
    16480
    15134
    ─────
    13460
    12972
    ─────
     4880
     4324
     ─────
      556 rem.
```

(5) 254)1160(.4566+ Ans.
```
    1016
    ────
    1440
    1270
    ────
    1700
    1524
    ────
    1760
    1524
    ────
     236 rem.
```

CASE 2.

(2)   Thus 2*R.* 4*P.*=84*P.*   1*A.*=160*P.*

     Then 160)840(.525   Ans.

```
          800

          400
          320

          800
          800
```

(3)   2*qr.* 2*na.*=10*na.*   And 1*yd.*=16*na.*

     Then 16)100(.625.   Ans.

```
           96

           40
           32

           80
           80
```

(4)   1*hr.*=60*min.*   And 60)5.00(.08333+   Ans

```
                              480

                              200
                              180

                              200
                              180

                              200
                              180

                               20 rem.
```

(5)   1*oz.*=480*grs.*

  Then 480)1000(.02083+ Ans.

```
         960

        4000
        3840

        1600
        1440

         160 rem.
```

(6) *2qts. 1pt.=5pts.*
     1hhd.=504pts. Then 504)5000(.00992+ Ans.
                                  4536
                                  ————
                                  4640
                                  4536
                                  ————
                                  1040
                                  1008
                                  ————
                                    32 rem.
                                  ————

CASE 3.

£		Day.		Gal.
(2) .1361		(3) .235	(4)	.42
20		24		4
————		————		——
s.2.7220		940	qt.1.68
12		470		2
————		————		——
d.8.6640		hrs.5.640	pt.1.36 Ans. 1 1.36
4	s.d.	60		
———— Ans. 2 8½		————		
qr.2.6560		min.38.400		
————		60		
		————	hrs. min. sec.
		sec.24.000 Ans. 5 33 24

qt. pt. heading appears above "Ans. 1 1.36"; *hrs. min. sec.* above "Ans. 5 33 24"

s.		Yd.		Acre.
(5) .253		(6) .436	(7)	9
12		4		4
———— d.		——		——
d.3.036 Ans.3.036	qr.1.744			r.3.6
		4		40
		——	qr. na.	—— ⸑R.P.
		na.2.976 Ans. 1 2	p.24.0 Ans. 3 24

RULE OF THREE IN DECIMALS.

EXAMPLES.

2) Thus 1.4*yd.* : 15.*yd.* :: 13*s.* : £6 19*s.* 3*d.* 1.71*qr.*
For 13×15=195. the dividend.
Then 195÷1.4=£6 19*s.* 3¼.71*qr.*

3) Thus 1*qr.* : 1*yd.* :: $2 34.5*cts.* : $9 38*cts.*
For 2. 34.5×4=$9 38*cts.* Ans.

4) First sold it for $108.30*cts.*
but paid for it   84.39.12—

gained on it $23.90.88

.Then 10.5*cwt.* : 1*cwt.* :: $23'90*cts.* 88*m.* : $2 27*cts.*
7*m.*+
For 23 .90 88÷10.5=$2 27*cts.* 7*m.* Ans.

5) Thus $20.8 : $12.6 :: 240*pie.* : 145.38*pie.*+
For 240×12.6=3024.0 which÷20.8=145.38*pie.*+
Ans.

6) Thus 3.5*oz.* : 5.2*oz.* :: 74.6*cts.* : $1 10*cts.* 8*m.*
For 5.2×74.6÷3.5=$1 10*cts.* 8*m.*

POSITION.

SINGLE POSITION.

EXAMPLES.

(2)   Suppose 162 in the box.

$$
\begin{array}{c|l}
\frac{1}{2} & 32.40 = \frac{1}{2} \\
\frac{1}{3} & 27.00 = \frac{1}{3} \\
\frac{1}{4} & 20.25 = \frac{1}{4} \\
\frac{1}{12} & 13.50 = \frac{1}{12}
\end{array}
$$

Result 93.15

Then $93 15*cts.* : $162 :: $690 : $1200. Ans.

(3) Suppose C.'s 40
+8
———
48=B.'s
+16
———
64=A.'s
48=B.'s
40=C.'s
———
152 result.
———

yrs   yrs.   yrs.
Then 152*yrs.* : { 64 ·: 133 : 56=A.'s }
{ 48 :: 133 : 42=B.'s } Ans.
{ 40 :: 133 : 35=C.'s }
———
133 proof.
———

(4) Suppose No. 3 cost $20
3
———
60=No. 2.
2
———
120=No. 1.
60
20
———
Result 200
———

$ $ $
Then 200 : { 120 :: 350 : 210=No. 1. }
{ 60 :: 350 : 105=No. 2. } Ans.
{ 20 :: 350 : 35=No. 3. }

M

Yrs.

(5)   Suppose 60
                      2
                   —————
                    120
                      3
                   —————
                  5)360
                   —————
                  3)72
                   —————
                    24 result.
                   —————

Then 24*yrs.* : 60*yrs.* :: 14*yrs.* : 35*yrs.*   Ans.

£

(6)   Thus suppose 40
                       $5\frac{3}{4}$
                    —————
                     200
                      20
                      10
                    —————

Int. for 1*yr.* $\begin{cases} £2|30 \\ 20 \\ \\ s.6|00 \end{cases}$

Then as £10 14*s.* 8*d.* : £201 5*s.* :: £40 : £750. Ans.

                    £   *s.*
     And $|\frac{1}{2}|$2   6
                           4 years.
                   —————————
Int. in 4 yrs. 9   4
Int. for 8 mo. $\left\{ |\frac{1}{2}| \begin{matrix} 1 & 3 \\ 0 & 7 & 8 \end{matrix} \right.$
                   —————————
   Whole int. 10   14   8
                   —————————

(7) Thus, suppose the cistern to hold 100 gallons.

Then $100 \div 45 min. = 2\frac{2}{9} gal.$ =the quantity which the first cock discharges in a minute.

And $100 \div 55 min. = 1\frac{9}{11} gal.$ the quantity which the second cock discharges in 1 min.

Then $100 \div 30 min. = 3\frac{1}{3} gal.$ =the quantity which the discharging cock discharges in 1 min. Consequently, $2\frac{2}{9} gal. + 1\frac{9}{11} gal. = 4\frac{4}{99} gal.$ the quantity which the cistern receives by both the first and second cocks in a minute. Then as $3\frac{1}{3} gals.$ run out in the same time, $4\frac{4}{99} gal. - 3\frac{1}{3} gal. = \frac{70}{99} gal.$ that the cistern gains in 1 min.

Then $\frac{70}{99} gal. : 100 gal. :: 1 min. : 2 hrs. 21 min. 25\frac{5}{7} sec.$ Ans.

DOUBLE POSITION.

(2) First suppose they received $276

$$276$$
$$2$$
$$\overline{}$$
$$3)552$$
$$184 = \text{what A. spent.}$$
$$+250$$
$$\overline{}$$
$$434 = \text{what B. spent.}$$
$$-276$$
$$\overline{}$$
$$158 \text{ B. was in debt every}$$
$$7 \qquad \text{year.}$$
$$\overline{}$$
$$1106 = \text{7 years' debt.}$$
$$-350$$
$$\overline{}$$
$$756 \text{ error too much.}$$

Again suppose the salary was 300 $

                   2

            3)600

               200=A. spent.
             +250

               450 B. spent.
             —300

B. was every year 150 in debt.

                 7

And in 7 years he was 1050 in debt.
               —350

               700 error too much.

Then $756 \times 300 = 226800$

$700 \times 276 = 193200$

Difference of errors   56)33600($600 the salary, $\frac{2}{3}$

                    336   of which=400

                           A.'s share, then

                    00   400+250=650

                           B.'s share. Ans.

(3)   First suppose 30 working days.

               1

             $30
           —10 that he forfeits.

Receives 20

            27 50

            7 50 error too little.

Again suppose 20 working days.

1

$20

Forfeits 15
___

Receives 5
    27 50
___

    22 50 error too little.
___

Then $2250 \times 30 = 67500$

$750 \times 20 = 15000$
___

Difference of errors 1500)52500(35 working days.

    4500
___

    7500

    7500
___

Therefore $50 - 35 = 15$ idle days.   Ans.

$

(4)   First suppose 10 cows = 160

    And 10 oxen = 240

    40 calves = 240
___

The whole 640

    −320
___

    320 error too much.

$

Again suppose 8 cows = 128

    And 8 oxen = 192

    And 32 calves = 192
___

The whole 512

    320
___

    192 error too much.

Then 320×8=2560
192×10=1920

Difference of errors 128)640(5cows 5oxen & 20calves.
640　　　　　Ans.

(5)　First suppose　　　　　Again suppose
Ft.　　　　　　　　　Ft.
No. 2=20　　　　　　No. 2=30

10=½　　　　　　15=½
15　　　　　　　15

25=No. 3.　　　　30=No. 3.
+15　　　　　　　+15

40　　　　　　　45=No. 2.
—20　　　　　　　—30

20 error too much.　15 error too much.

Then 20×30=600
15×20=300

Difference of errors 5)300

60=No. 2, then 60—15=45=
—　　　　No. 3.

And then we have No. 1=15, No. 2=60, and No.
3=35, which added together=120ft. the length of
the pole.　Ans.

(6) Thus first suppose the whole property to have been worth £

£ 396	Again suppose £ 432
$198=\frac{1}{2}$	$216=\frac{1}{2}$
—40	—40
158=A.'s share.	176=A.'s
$132=\frac{1}{3}$	$144=\frac{1}{3}$
+12	+12
144=B.'s share.	156=B.'s
—80	—80
60=C.'s share.	76=C.'s
144	156
158	176
366 sum.	408 sum.
396	432
30 error of defect.	24 error of defect.

Then 432×30=12960
396×24= 9504

Difference of errors 6)3456

£576 Ans.

£
Then 576÷2—40=248 A.'s share.
204÷3+12=204 B.'s do.
204—80=124 C.'s do.

£576 proof.

(7)   First suppose each boy received   £ 3

                                         2

                                  6 = share of each

                                  3   . woman.

                                18 = share of each

                                         man.

$$
\begin{aligned}
&\text{And } 19\times3 = 57\\
&\phantom{\text{And }}11\times6 = 66\\
&\phantom{\text{And }}7\times18 = 126\\
&\phantom{\text{And }19\times6=}\overline{}\\
&\phantom{\text{And }19\times6=}249\\
&\phantom{\text{And }19\times6=}172\ \ 19\ \ 4\tfrac{1}{4}
\end{aligned}
$$

                                 76   0 $7\tfrac{3}{4}$ error of excess.

Again suppose each boy received   £ 1

                                         2

                                  2 share of each woman.

                                  3

                                  6 share of each man.

$$
\begin{aligned}
&\text{And } 19\times1 = 19\\
&\phantom{\text{And }}11\times2 = 22\\
&\phantom{\text{And }}7\times6 = 42\\
&\phantom{\text{And }19\times6=}\overline{}\\
&\phantom{\text{And }19\times6=}83\\
&\phantom{\text{And }19\times6=}172\ \ 19\ \ 4\tfrac{1}{4}
\end{aligned}
$$

                                 89 19 $4\tfrac{1}{4}$ error of defect.

$$
\begin{array}{llll}
& £ & s. & d. & & £ & s. & d. \\
\text{Now } 89 & 19 & 4\frac{1}{4} \times 3 = & 269 & 18 & 0\frac{3}{4} \\
76 & 0 & 7\frac{3}{4} \times 1 = & 76 & 0 & 7\frac{3}{4}
\end{array}
$$

$$345 \quad 18 \quad 8\frac{1}{2}$$

Which ÷ 166 sum of errors = £2 1s. 8d. + = each boy's share, which × 2 = £4 3s. 4¼d. + = each woman's share, which × 3 = £12 10s. 0¾d. + = each man's share. Ans.

INVOLUTION, OR THE RAISING OF POWERS.

EXAMPLES.

(2) 14×14×14=2744. Ans.
(3) 2.8×2.8×2.8×2.8×2.8×2.8=481.890304. Ans.
(4) .263×.263×.263=.018191447. Ans.
(5) $\frac{1}{4}\times\frac{1}{4}\times\frac{1}{4}\times\frac{1}{4}\times\frac{1}{4}\times\frac{1}{4}\times\frac{1}{4}\times\frac{1}{4}=\frac{1}{65536}$. Ans.
(6) 401×401×401×401=25856961601. Ans.

EVOLUTION, OR THE EXTRACTING OF ROOTS.

SQUARE ROOT.
EXAMPLES.

(2) 39375655(6275 Ans.   (3) 1486.179010(38.55. Ans.
    36   9

    122)337   68)586
       244   544

    1247)9356   765)4217
        8729   3825

    12545)62755   7705)39290
         62725   38525

    Rem. 30   Rem. 76510

(4) 96385163(9817 Ans.   (5) .0001324960(.01151 Ans.
     81   1

188)1538   21)32
    1504   21

1961)3451   225)1149
     1961   1125

19627)149063   2301)2460
      137389   2301

Rem. 11674   Rem. 159

(6)   18.362147(4.285   Ans.
         16

    82)236
       164

    848)7221
        6784

    8565)43747
         42825

    Rem. 922

(7) $\frac{3450}{3200}=\frac{49}{64}$ whose square root is $\frac{7}{8}$,   Ans.

(8) 36)$\frac{1296}{1764}=\frac{36}{49}$ whose square root is $\frac{6}{7}$.   Ans.

(9) 500)3200($\sqrt{}$.64(.8 Ans.
     3000   64

        2000
        2000

(10) $50 \times 64 + 49 = 3\frac{249}{64}$.

   Then 3249($\frac{57}{8} = 7\frac{1}{8}$. Ans.
        25

   107).749
        749

And $\sqrt{}$.64(.8 denominator.
       64

(11) $30 \times 100 + 25 = 30.25$

   Then 30.25(5.5 = $5\frac{5}{10}$. Ans.
        25

   105)525
       525

(12) 1296(36 Ans.
   $3 \times 3 = 9$

        66)396
           396

(13) 169(13 Ans.
      1

   23)69
      69

(14) 3097600(1760yds. = 1mile.
      1   Ans.

   27)209
      189

   346)2076
       2076

          00

(15)   Thus $15 \times 15 = 225$

$24 \times 24 = 576$

$\sqrt{801}(28.3\text{ft. Ans.}$

4

48)401

384

563)1700

1689

Rem. 11

(16) $212 \times 212 = 44944 ft.$

And $20 yds. = 60 \times 60 = 3600 ft.$

41344(203.332$ft.$ Ans.

$2 \times 2 = 4$

403)1344

1209

4063)13500

12189

40663)131100

121989

406662)911100

813324

Rem. 97776

CUBE ROOT.

(2)   7532641(196.02 Ans.
                                        1
                                        ——
{ Defec. div. and squ. of 9=381 6532
{ +270=com. divisor   =651 5859

{ Def. div, and squ. of 6=108336 673641
{ +3420=com. div.   =111756 670536

    Defective divisor 115248   3105000

{ Def. di. & sq. of .02=1152480004)3105000000
{ +11760=com. div.=1152491764)2304983528

                        Rem. 800016472

(3)   12.113847500(2.296 Ans.
                    2×2×2=8

{ Def. div. and squ. of 2=1204)4113
{ +120=com. divisor =1324)2648

{ Def. div. & sq. of 9=145281)1465847
{ +5940=com. di. =151221)1360989

{ Def. di. & sq. of 6=15732336)104858500
{ +61830=co. di. =15773556).94641336

                Rem. 10217164

(4)   5382674(175.2   Ans.
                           1

{ Defec. div. and square of 7=349)4382
{ +210=complete divisor   =559)3913

{ Defec. div. & square of 5=86725)469674
{ +2550=complete divisor=89275)446375

{ Defec. div. and squ. of 2=9187504)23299060
{ +10500=com. divisor   =9198004)18396008

                 Rem. 4902992

(5)   .378621350(.723. Ans.
           $7 \times 7 \times 7 = 343$

{ Defec. div. & sq. of 2=14704)35621
{ +420=com. divisor =15124)30248

{ Def. div. & squ. of 3=1555209)5373350
{ +6480=com. divi. =1561689)4685067

                Rem. 688283

(6)   46.295363543(3.590 Ans.
        $3 \times 3 \times 3 = 27$

{ Def. div. & sq. of 5=2725)19295
{ +450=com. divi. =3175)15875

{ Def. div. & sq. of 9=367581)3420363
{ +9450=com. di. =377031)3396279

     Defective divisor 128881)24084543

(7) Thus $4)\frac{208}{1036}=\frac{52}{259}$, which reduced to a decimal=
.2007722

Then .200772200(.585 Ans.
125

$\begin{cases} \text{Defec. divisor \& squ. of } 8=7564)75772 \\ +200=\text{complete divisor}=8764)70112 \end{cases}$

$\begin{cases} \text{Defec. div. and sq. of } 5=1009225)5660200 \\ +8700=\text{comp. divisor}=1017925)5089625 \end{cases}$

570575 Rem.

(8) Thus $\sqrt[3]{36.\frac{26}{30}}=\sqrt[3]{36.866666}+(3.32$ Ans.
$3\times3\times3=27$

$\begin{cases} \text{Defec. div. \& squ. of } 3=2709)9866 \\ +270=\text{complete divi.}=2979)8937 \end{cases}$

$\begin{cases} \text{Defec. div. \& squ. of } 2=326704)929666 \\ +198=\text{com. divisor}=329684)657368 \end{cases}$

Rem. 272298

ALLIGATION.

CASE I.

	cwt.	$ cts.	$ cts.
(2)	2 at 25	= 50 00	
	4	20 50	= 82 00
	7	18 62½	=130 37½
	13		$262 37½

Then as 13cwt. : 1cwt. :: $262 37½ cts. : $20 18¼ cts. Ans.

CASE 2.

cts.

$$(2) \quad \text{Mean rate } 50 \left\{ \begin{array}{l} 34 \\ 42 \\ 86 \\ 110 \end{array} \right. \begin{array}{l} =36 \text{ at } 34 \text{ cts.} \\ =60 \text{ at } 42 \text{ cts.} \\ =16 \text{ at } 86 \text{ cts.} \\ = 8 \text{ at } 110 \text{ cts.} \end{array} \left. \right\} \text{Ans.}$$

CASE 3.

cts.

$$(2) \quad \text{Mean rate } 92 \left\{ \begin{array}{l} 75 \\ 86 \\ 94 \\ 105 \end{array} \right. \begin{array}{l} = 2 \\ =13 \\ =17 \\ = 6 \end{array}$$

lbs.

Then 2 : 6 :: 13 : 39 at 86 cts. ⎫
    2 : 6 :: 17 : 51 at 94 cts. ⎬ Ans.
    2 : 6 :: 6 : 18 at 105 cts. ⎭

CASE 4.

cts.

$$(2) \quad \text{Mean rate } 145 . \left\{ \begin{array}{l} 130 \\ 160 \\ 180 \end{array} \right. \begin{array}{l} =15+35=50 \\ = \quad\quad\quad 15 \\ = \quad\quad\quad 15 \end{array}$$

                         80 sum of differ.

Then as 80 : 50 :: 32 : 20 at 130 cts. ⎫
       80 : 15 :: 32 : 6 at 160 cts. ⎬ Ans.
       80 : 15 :: 32 : 6 at 180 cts. ⎭

ARITHMETICAL PROGRESSION.

CASE L

EXAMPLES.

(2) Thus 40—1=39
　　　　2 com. dif.
　　　　———
　　　　78
　　　　2=1st term.
　　　　———
　　　　80
　　　　2=1st term.
　　　　———
　　　　82 sum.
　　　　40
　　　　———
　　　2)3280
　　　　———
　　　$16.40 Ans.

(3) 10—1=9
　　　　　4 com. dif.
　　　　　———
　　　　　36
　　　　+10=1st term.
　　　　　———
1st Ans. 46 last term.
　　　　+10
　　　　———
　　　　56
　　　　10
　　　　———
　　　2)560
　　　　———
　　　　280　2dAns.

(4)　　　75—1=74
　　　　　2 common difference.
　　　　　———
　　　　　148
　　　　+6=1st term.
　　　　　———
　　　$1.54 for the last.　1st Ans.
　　　　6=1st term.
　　　　———
　　　　160 sum.
　　　　75
　　　　———
　　　　800
　　　1120
　　　　———
　　　2)12000
　　　　———
　　　$60 00 in the whole. Ans.

CASE 2.

(2)   Thus 175
                        —21
                        ———
8—1=7)154
         ————

            $22 common difference.
            ——

And 175+21=196 sum of extremes.
                    8 number of terms.
                    ——

        2)1568
        ————

        784 whole sum.
        ——

Lastly 21+22= 43=2d payment.
       43+22= 65=3d
       65+22= 87=4th
       87+22=109=5th
      109+22=131=6th
      131+22=153=7th
      153+22=175=8th
                ——

        763
         21=1st payment.
        ——

    $784 proof.
    ——

(3) Thus 49   Then 49+4=53 sum of extremes.
        —4   10 number of terms.
        ——   ——
10—1=9)45   2)530
       ——   ————
    5 com. dif. Received $2.65 Ans.
    ——

GEOMETRICAL PROGRESSION

EXAMPLES.

(2) Thus power 1 2 3 4
        Ratio 3 9 27 81

                        27 3d power

                        567
                        162

                        2187=7th power.
                           5=1st term.

1st Ans. 10935=last term.
                3 ratio.

                32805
                   —5=1st term.

Ratio less 1=2)32800

                £16400 Ans. 2d

(3) Thus power 1 2 3 4 5 6 7 8 9
        Ratio 2 4 8 16 32 64 128 256 512

                                512

                                1024
                                512
                                2560

                                262144=18th p.
                                     4=2d do.

                                1048576=20th p.
                                      1 1st term.

                                1048576=last term.
                                      2 ratio.

                                2097152
                                      1=1st term.

Ratio less 1=1)2097151

                Ans. $20971.51cts.

COMPOUND INTEREST BY DECIMALS.

EXAMPLES.

(2) Thus, tabular number 1.2155062
                                750

                           607753100
                            85085434

                           911.6296500
Amount of £1 for 6*mo.*   1.024695 from table first.

                           45581482500
                           82046668500
                           54097779000
                          . 36465186000
                           18232593000
                           91162965000

                      £934.1423442067500
                                       20

                      *s.*2.8468841350000
                                       12

                      *d.*10.1626096200000

                          £   *s.* *d.*
                Amount 934   2   10+
                Principal 750 0   0

                Interest 184 2   10+ Ans.

CASE 2.

(1) Thus £695 13*s.* 9*d.*=695.6875£.
Then from tab. II. 1.2762815)695.68750(545£ 1*s.*
9*d.*+ Ans.

(2) Thus £260 5*s.* 3*d.*=260.2625£ which÷by 1.191016
from table II.=£218 10*s.* 5*d.*+ Ans.

—••○◉○••—

ANNUITIES AT COMPOUND
INTEREST.

CASE 1.

(2) The number from table III.=5.637093
$$\underline{\qquad\qquad 200\text{=annuity.}}$$

Amount for yearly payments=1127.4186 which ×
1.014781 proper number for $\frac{1}{2}$ yearly payment from
table V.=$1144 08 2*m.*+ Ans.

CASE 2.

(2) Thus, the num. from tab. IV.=4.21236
$$\underline{\qquad\qquad £70\text{ annuity.}}$$

$294 86 52 Ans. for yearly
payments.
Then $294.8652 × 1.014781 from table V. =
$299.22.3+*mills.* Ans. for $\frac{1}{2}$ yearly payments.

And 294.8652×1.022257 for quarterly payments
from the same table=$301.42.8+*mills.* Ans. for
quarterly payments.

ANNUITIES IN REVERSION.

(2) Thus 9+4=13*yrs.*=9.98565 table IV.
4 do.=3.62989—

$$6.35576$$
$$120$$

$$1271152$$
$$635576$$

$762.69.1.2*m.* Ans.

PERPETUITIES AT COMPOUND INTEREST.

(2) Thus, ratio—1=1.06—1=.06)260.00

$4333.33.3*m.*+ Ans.

COMBINATION.

EXAMPLES.

(2) Thus $20 \times 19 \times 18 \times 17 \times 16 \times 15 \times 14 \times 13 \times 12 \times 1 = 1 \times 2 \times 3 \times 4 \times 5 \times 6 \times 7 \times 8 \times 9 \times 10 =$

$$\frac{670442572800}{} = 184756 \quad \text{Ans.}$$

PERMUTATION.

EXAMPLES.

(2)   Thus $1\times2\times3\times4\times5\times6\times7\times8\times9\times10\times11\times12=$
479001600 number of changes.
15 seconds.

$$\begin{array}{r} 2395008000 \\ 479001600 \end{array}$$

6|0)718502400|0 sec.

6|0)11975040|0 min.

$365\frac{1}{4}d.=$8766 hrs.)1995840(227 yrs. 248 days. 6 hrs.
Ans.

DUODECIMALS.

ADDITION OF DUODECIMALS.

EXAMPLES.

	Ft.	in.	"	"'	""
(1)	10	5	6	11	6
	15	9	5	2	10
	18	4	1	7	9
	12	8	6	5	7
Ans.	57	3	8	3	8

	Ft.	in.	"	"'	""
(2)	37	8	10	6	9
	43	11	2	4	7
	19	7	5	3	8
	18	4	1	7	2
Ans.	119	7	7	10	2

		Ft.	in.	"
(3)		16	8	0
		14	6	0
		17	9	2
	Ans.	48	11	2

SUBTRACTION OF DUODECIMALS.

EXAMPLES.

		Ft.	in.	"	'''	''''
(1)	From	38	8	4	7	5
	Take	15	11	6	9	3
	Ans.	22	8	10	2	2

		Ft.	in.	"	'''	''''
(2)	From	720	3	8	1	6
	Take	13	9	4	7	10
	Ans.	706	6	3	5	8

		Ft.	in.	"	'''	''''
(3)	From	475	7	2	0	0
	Take	81	2	5	10	6
	Ans.	394	4	8	1	6

MULTIPLICATION OF DUODECIMALS.

CASE I.

EXAMPLES.

	Ft.	in.	
(2)	54	10	
		5	7
	31	11	10
	274	2	
Ans.	306	1	10

	Ft.	in.	"
(3)	6	9	3
		3	5
	2	9	10	3
	20	3	9
Ans.	23	1	7	3

CASE 2.

(2)

		Ft.	in.	"		
6	½	81	10	4		
				$7 \times 2 = 14$		

```
              573   0   4
                        2
            ───────────────
             1146   0   8
       1        40  11   2
  1  1/6          6   9  10   4
  4" 3            2   3   3   5  4
  1  1/4          6   9  10   4
            ───────────────────────
      9)1196   7   9   7   8
    ─────────────────────────────
  Yds. 132   8   7   9   7   8  Ans.
```

(3)

in.		Ft.	in.	"	'''
4	1/3	2	5	7	2

```
  1   1/4   0   9  10   4   8
  3"  1/4       2   5   7   2
  4   1/4           7   4   9   6
  6''' 1/8          1   2   9   7
  4'''' 1 8                 9  10   4   8
  1   1/4                   2   5   7   2
       ─────────────────────────────────────
       1   1   0   8   5   4  11  10  contents of 1 sh.
                              10×10×10=1000
       ─────────────────────────────────────
      10  10   7   0   6   1  10   4
                                  10
       ─────────────────────────────────────
     108   9  10   5   1   6   7   4
                                  10
       ─────────────────────────────────────
 sq. ft. 1088  2   8   3   3   6   1   4  Ans.
```

PROMISCUOUS EXAMPLES.

(1)

Thus A.'s 25 years.
+15
—
B.'s 40 years.
+12
—
C.'s 52 years. Ans.
—

(2)

	$ cts.	$ cts.	

Thus 220 50÷5=44 10 A.'s own share.
220 50÷6=36 75 B.'s do.
————
80 85 sum.
220 50
————
139 65=C.'s own share.
————

$ cts. $ cts. m.
Then 36 75÷2=18 37 5=½ B.'s share.
44 10
————
62 47 5=A.'s last share.
————

$ cts. m.
And 18 37 5
139 65
————
Ans. 158 02 5=C.'s last share.
————

(3) $100—7½ : $100 :: $56 25cts. : $60 81cts. 5m.+25.
For 5625×100=562500 the dividend.
And 100—7½=92½ the divisor.
Then 562500÷92½=$60 81cts. 5m.+25. Ans.

(4) Thus B. gains 2 miles per hour.
Then as 2m. : 50m. :: 1hr. : 25hrs. 1st Ans.
Now as B. went at the rate of 10 miles per hour for
25 hours, 10×25=250 miles, the 2d Ans.

(5) Thus $\frac{5}{20}=\frac{1}{4}$)750

187 50 whole price of the damaged.
100   loss.

87 50 what it sold for.

Then $1 25cts. : $87 50cts. :: 1yd. : 70yds.=quanti-
ty damaged.
And 70×4=280yds. the whole quantity.
        70

210 undamaged.

And $750 00cts. cost.
        87 50 received for the damaged.

210yds. : $662 50 :: 1 : $3 15¼cts.+ Ans.

5) Thus 1000—1=999 number of terms—1.

1 ft. common difference.

———

999

2 ft. first term.

———

1001 last term.

2 .

———

1003 sum of the terms.

1000

2)1003000

———

3)501500 ft.

220)167166+2 ft.

———

8)759+186 yds.

94+7*fur.* 186*yds.* 2*ft.* Ans.

———

7) Thus admit the wall to contain 3600 feet.

Then 20)3600(180 feet raised in a day by A. B. & C.

24)3600(150   B. C. & D.

30)3600(120   C. D. & A.

36)3600(100   A. B. & D.

———

3)550

———

183⅓ feet per day by altogether.

———

Then 183⅓   And 183⅓

B. C. & D. 150   C. D. & A. 120

———

A. 33⅓   B. 63⅓

And $183\frac{1}{3}$ And $183\frac{1}{3}$
A. B. & D. 100   A. B. & C. 180

C. $83\frac{1}{3}$ D. $3\frac{1}{3}$

days.

Then, feet per day by A. $33\frac{1}{3}$)3600(108 for A. to do it in.
    do.   by B. $63\frac{1}{3}$)3600($56\frac{16}{19}$ B.   do.
    do.   by C. $83\frac{1}{3}$)3600($43\frac{1}{5}$ C.   do.
    do.   by D. $3\frac{1}{3}$)3600(1080   D.   do.
And $183\frac{1}{3}$)3600($19\frac{7}{11}$ days all working together. Ans.

                  d.   *d.*
(8) Thus 4 crowns at 146 each=584
        3 dolls.   108   =324
        2 ducats   136   =272

                      1180*d.* sum.

And £1055 15*s.*=253380*d.*
                 d.   *d.*   *d.*
Then   { 584 : 125402÷146=$858\frac{62}{13}$*cr.* }
1180 : 253380 :: { 324 : 69572÷108=$644\frac{5}{27}$*$.* } Ans.
                    { 272 : 58406÷136=$429\frac{31}{68}$*duc.* }

(9) Thus 9*m.* : 21*m.* :: $332 50*cts.* : $775 $83\frac{1}{3}$*cts.*   Ans.
For 33250×21=698250 the dividend.
And 9=the divisor.
Then 698250÷9=$775 $83\frac{1}{3}$*cts.*

(10)   Thus 12
                   4

            16*yrs.*=10.83777 Table IV.
Time of reversion 12   = 8.86325   do.

                 1.97452 difference.
                 720.25 annuity.

                  987260
                  394904
               3949040
             1382164

             $1422.1480300

     Or $1422 14*cts.* 8*m.*+   Ans.

(11) 3150 gigs÷7×5= $ *cts.*
    2250 wagons wh.   } 135 00 for the wagons.
    ×6*cts.*=
    3150 gigs÷3×5=
    5250 footmen wh.   } 52 50 for footmen.
    ×1*ct.*=
    5250 footmen÷6×4
    =3500 horsemen   } 70 00 for horsemen.
    which×2*cts.*=
    3150 gigs at 4*cts.* per   } 126 00 for gigs.
    gig=

    Amount of toll 383 50 Ans.

(12) Thus 15*gals.* in 3*min.*=5*gals.* per min. that run in.
    And 20÷5=4*gals.* that run out in a min. Con-
    sequently, the gain is 5—4=1*gal.* per min. which
    is 60*gals.* per hour.
    Then 110—60=50*gals.* yet to run in.
    Then 5*gals.* : 50*gals.* :: 1*min.* : 10*min.*   Ans.

(13) Thus 264
          6
   mo. ————

6|½|15 84 Int. for 1 year.

3|½| 7 92
   | 3 96

   11 88 Int. for 9 months.
   264 00
   30 00 profit.
   ————
   $305 88 for the whole.
   ————

          lbs. $ cts. m.
Then 28cwt.=3130)30588(0 9 7+ Ans.
           28224
           ————
           23640
           21952
           ————
   Rem.   1688
   ————

(14) Thus, the proportions are A. 4, B. 5, C. 3=12.
                    $
Then 12 : 780 :: { 4 : 260 A.'s share of profit } 1st
                 { 5 : 325 B.'s   do.   } Ans.
                 { 3 : 195 C.'s   do.   }

   $780 proof.
   ————

          $ mo.
Then 260×5=1300
     325×7=2275
     195×9=1755
     ————
     5330

Again 5330 : 5762 :: $\left\{ \begin{array}{l} 1300 : 1405 \ 36 \ \text{A.'s stock.} \\ 2275 : 2459 \ 39 \ \text{B.'s} \\ 1755 : 1897 \ 25 \ \text{C.'s} \end{array} \right.$

$5762 00 proof.

Now 2459 39
2087 00 B. received.

372 39 B.'s loss of stock.
And 325 00   do.   of gain.

Ans. $697 39 A. & C. would gain.

(15) $100 + 5\frac{3}{4} = 105 \ 75$ $ cts. m.
Then 105 75 : 100 :: 1000 : 945 62 6 cost C.
    20 75 0 less.

$924 87 6 cost B.

Again 100
—5 50

94 50 : 100 :: $924 87cts. 6m. : $978 70cts.
4m. that the whole cost A. which÷20hhds.=$48
93cts. 5m.+ per hhd. Ans.

(16) $10 \times 11 = 110$ sold for.
    $1 \times 7 = 70$ worth.

$40 gain of A.

| $ cts. m. | | $ cts. |
And 110÷3= 36 66 6+ paid cash.   5 25
    110 00 0   4 50

$73 33 3 to pay in paper. $0 75 B. gains.

Then 450 : 75 :: 73 33 3 : $12 22cts. 2m. gain of
B. $40—$12.22.2=$27.77.8. Ans.

(17)   Thus 21—14=7 years to be of age.
                Then $1300
                       6
                   ─────
                    7800 int. for first year.
                  1300
                   ─────
                    1278 amount—100.
                       6
                   ─────
                    7668 int. second year.
                  1278
                   ─────
                   125468 amount—100.
                       6
                   ─────
                    752808 int. third year.
                  125468
                   ─────
                   12299608 amount—100.
                       6
                   ─────
                    73797648 int. fourth year.
                  12299608
                   ─────
                   12037584 amount—100.
                       6
                   ─────
                    72225504 int. fifth year.
                  12037584
                   ─────
                   11759839 amount—100.
                       6
                   ─────
                    70559034 int. sixth year.
                  11759839
                   ─────
                   11465429 amount—100.
                       6
                   ─────
                    68792574 int. seventh year.
                  11465429
                   ─────
        $1115.33.54m. amount—100.   Ans.

Another solution:

First, $1.06^7 = 1.5036302$. See table II. Arithmetic.

And $1.5036302 \times 1300 = 1954.719$ Amount at Compound Interest.

Also, $8.393837 \times 100 = 839.383$.   Amount of \$100 Annuity for 7 years, table III.

Hence \$1954.719—\$839.383=\$1115 33*cts.* 5*m.* Ans.

(18)

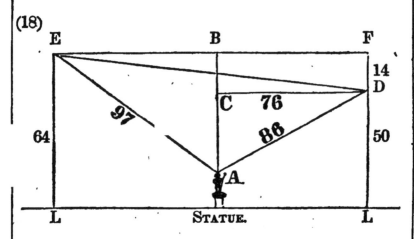

Thus, referring to the above figure.

A B is a perpendicular line erected on the centre of the statue's base, which forms the side A C of the right angle A C D; and the other two sides, A D 86 and C D 76, are given to find the length of the side A C.

Now $76^2 = 5776$ & $86^2 = 7396$
$$-5776$$

$\sqrt{1620}$ diff.$(40.2+ = \text{A C}$

Then 40.2+14 the difference between the columns
=54.2 the whole length of A B.   Then 54.2²=
2937.64 & 97²=

that is A E=9409
            —2937.64

$\sqrt{6471.36}$=(80.44+ for E B
            +76   that is B F

14=D E   156.44=E B F
14   156.44

56   62576
14   62576
                                93864
195   78220
                                15644

                            24473.4736
                            196

            $\sqrt[2]{24669.4736}$=157ft.   Ans.

NOTE.—This solution supposes the statue to be lower than the
columns: admitting it to be higher, the operation will, of course, be
different; but may readily be performed from the one here given.

(19)   1sec. : 47sec. :: 1150ft. : 54050ft.   Ans.

(20)   15m. 7fur.=83820ft.
        Then 1150ft. : 83820ft. :: 1sec. : 1m. 12$\frac{102}{115}$sec. Ans

(21) First suppose ½ of 8.2245 in. to be gold.

4.11225=½	4.11225 in. of sil.
10.36	5.85

2467350	2056125
1233675	3289800
4112250	2056125

42.6029100oz. g.	24.0566625oz. sil.
24.0566625	

66.6595725
63

3.6595725 error of excess.

Again suppose ⅓ of 8.2245 in. to be gold, the rest silver.

2.7415=⅓	5.4830=silver.
10.36	5.85

164490	274150
82245	438640
274150	274150

28.401940oz.	32.075550oz. sil.
32.075550	

60.477490
63.

2.522510 error of defect.

[See *following page.*

Then 3.6595725 × 2.7415 = 10.0327180875
And   2.522510 × 4.11225 = 10.3731917475

And 2.522510 + 3.6595725 = 6.1820625) 20.4059097625 (3.3008148 inches of gold.
18.5462475

185966225
185462475

5037750625
494566600

91840250
61820625

300194250
247283300

529109500
494566600

34542900 rem.

Then 3.3008148 × 10.36 = 34.196441328 ounces of gold, and the rest, which is 28.803558672 ounces silver. Ans

P

Another solution:

  oz. *oz.* *oz.*

First $63 \div 8.2245 = 7.66$ weight of a cubic inch of the mixture.

Then 7.66 $\begin{cases} 10.36) = 1.81 \text{ proportional } bulk \text{ of gold.} \\ 5.85) = 2.7 \text{ proportional } bulk \text{ of silver.} \end{cases}$

Also $1.81 \times 10.36 = 18.7516$ proportional *weight* of gold.
And $2.7 \times 5.85 = 15.795$ proportional *weight* of silver.

    34.5466 sum.

         oz.

Hence $34.5466 : 18.7516 :: 63 : 34.19587$ gold $\Big\}$ **Ans.**
And $34.5466 : 15.795 \ :: 63 : 28.80356$ silver.

     · Proof 62.99943

(22) Thus $7lbs.$ beef at $5\frac{3}{4}cts. = 40\frac{1}{4}cts.$
    5 bread at 6 $= 30$
  Then $40\frac{1}{4}cts. : \$34 \ 50cts. :: 30cts. : \$25 \ 71cts. \ 4m. +$
  Ans.

(23) Thus $\frac{4}{9}$ of $\frac{5}{7}$ of $\frac{362}{463} = \frac{7240}{29169}$.
  Then $1 - \frac{7240}{29169} = \frac{21929}{29169}$. Ans.

(24) $\$$
       1000
        6

      $60|00$ int. for 1 year.
        8

     $\$480$ int. for 8 years.

Then 8 years.
    6 per cent.

    ——

    48
   100

    ——

148 amt. of $100 for 8 years. at 5 per cent.

     $ $ $ $ cts. m.
Then 148 : 100 :: 1000 : 675 67 5 the present worth.
                1000 00 0

          $324 32 5 discount.
          480 00 0 interest.

     Ans. $155 67 5 difference.

(25) $\sqrt{32}=5.656+$
         $\sqrt{24}=4.9$

        10.556 sum.
     $\sqrt[3]{67}=4.06+$

    Ans. 6.496 difference.

(26)   Thus $100 : 105\frac{1}{2}$ :: $2450 : $2587 20cts.   Ans.

(27)   Thus the amount of $500 75cts. for 9 months at 6 per cent.=$533 28cts. 4m.

cts. $ cts.

And $5064 \times 2\frac{1}{2} = 126$ 60 price of the boards.
  $140 \times 13 = 18$ 20 do. tallow.

     144 80 amt.
     523 28 4

    $378 48 4 to receive in flax-seed.

Then as $92\frac{1}{2}cts.$: $378 $48cts.$ $4m.$ :: $1bu.$: $409\frac{150}{925}bu.$ Ans.

(28)   9 yrs.$=36$ qrs. the sum of terms.
     -1

     35
     3 common difference.

     105
     $+6=$1st term.

     111 last term.
     6$=$1st term.

     117 sum.
     $\times 36$ number of terms.

     702
     351

    2)4212

    $21.06cts.$ due him. Ans.

(29) Thus 5*yrs.* —2¼*yrs.*=2¾*yrs.*
Then $1.06 \times 1.06 \times 1.045 = 1.174162$ divisor.
And $2363.3875 \div 1.174162 = \$2012$ 82*cts.* 9*m.* Ans

(30) Thus, from January 14th, 1802, till July 5th, 1807
inclusive=5 years 173 days. And the amount o
$1854.69 for that time at 5 per cent. per annum=
$2362.3161
285.   paid off.
_____

2077.3161 second bond.
4¾
_____

83092644
10386580
5193290
_____

98.67.2514 int. of the 2d bond for 1 yr.
_____

Then 98672514 : 365 : : 52.65 : 194 days the tim
of the second bond.

Now 2077.3161
52.65   interest.
_____

2129.9661 amount.
102.43   paid off.
_____

2027.5361 3d bond.
_____

Which was out from January 12, 1808, till October 26th, 1813, which is 5.789 years.

$2497.0323 last amount.
2027.5361 last bond.
_____
469.4962 gained on the last bond, which was out 5.789 years, and this bond inclusive to the time=11737.4064829.

Then 11737.4064829 : 469.4962 : : 100 : 4 per cent. Ans.

(31) First suppose 10 horses at $50=$500
                   20 cows   20=400
                   60 sheep   4=240

                          $1140 sum.
                           456
                          _____
                           684 error of excess.

Again suppose 8 horses at $50=$400
              16 cows   20=320
              48 sheep   4=192

                      $912 sum.
                       456
                      _____
                       456 error of excess.

Then $684 \times 8 = 5472$
$456 \times 10 = 4560$
___

Difference of errors = 228)912(4 horses.
912
___

For 4 horses at $50 = 200
8 cows   20 = 160 } Ans.
24 sheep   4 = 96
___
$456 proof.
___

Another solution:

First $50 price of each horse.
$20 \times 2 = 40$ price of cows for each horse.
$4 \times 6 = 24$ price of sheep for each horse.
___
114)456(4 number of horses.
456
___

Then 4 horses at $50 = 200
$4 \times 2 = 8$ cows   20 = 160
And $8 \times 3 = 24$ sheep   4 = 96
___
456 proof.
___

32)   Thus $\begin{cases} 16 \\ 17 \\ 24 \end{cases}$
       Mean rate 19 $\begin{cases} 16)= & 5 \\ 17)= & 5 \\ 24)=3+2= & 5 \end{cases}$

        oz.   *oz.*

Then as 5 : 10 :: $\begin{cases} 5 : 10 \text{ of 17 carats fine.} \\ 5 : 10 \text{ of 24 carats fine.} \end{cases}$ Ans.

33) £100 : £120 :: £230 5*s.* : £276 6*s.* the amount in
    sterling.
    Then as £1 : £276 6*s.* :: $4 44*cts.* 4*m.* : $1227
    87*cts.* 7*m.*+ Ans.

34) Thus $\frac{37}{170}+\frac{1}{2}=\frac{518}{680}$, and $\frac{518}{680}$ subtracted from $1=\frac{162}{680}$
    =the 27 feet.
    Then $\frac{162}{680}$: 27*ft.* :: 1 : 113*ft.* 4*in.* Ans.

(35) $7 : 56$\frac{1}{4}$*cts.* :: $400 : $32 14$\frac{2}{7}$*cts.* Ans.

(36)   Thus 30
                +96
                ———
             126 sum.
               25 number of terms.
             ———
             630
             252
             ———
          2)3150
             ———
         $15.75 Ans.

(37) Thus 4 : 9 :: 47 : 105.75 the greater number.

47

152.75 sum.
58.75 difference.

76375
106925
122200
76375

Product 8974.0625 Ans.

THE END.

Lightning Source UK Ltd.
Milton Keynes UK
UKHW040636250219
337804UK00008B/1400/P